FERTILISING THE FUTURE

Bharat's March Towards Fertiliser Self-Sufficiency

FERTILISING THE FUTURE

Bharat's March Towards Fertiliser Self-Sufficiency

Dr. Mansukh Mandaviya

RUPA

Published by
Rupa Publications India Pvt. Ltd 2024
7/16, Ansari Road, Daryaganj
New Delhi 110002

Sales centres:
Bengaluru Chennai
Hyderabad Jaipur Kathmandu
Kolkata Mumbai Prayagraj

P-ISBN: 978-93-6156-974-6
E-ISBN: 978-93-6156-429-1

First impression 2024

10 9 8 7 6 5 4 3 2 1

Printed in India

Inspiration

Our Prime Minister Shri Narendra Modi ji,
who has dedicated every moment of his life
to the service of Bharat Mata

Gratitude

Crores of farmers, officials, employees, people from the Government and the private sector, all the stakeholders, including the operators of Pradhan Mantri Kisan Samriddhi Kendra, engaged at different levels in building "self-reliant Bharat" in the field of fertilisers.

Dedication

To crores of farmers of the country

Contents

APPENDICES

"PMKSK is a hub connecting with the farmers and helping them in every way"

– Shri Narendra Modi, Prime Minister

(English rendering of PM's speech at the inauguration of PM-KISAN Samman Sammelan on October 17, 2022 in New Delhi.)

Bharat Mata ki – Jai!
Bharat Mata ki – Jai!
Bharat Mata ki – Jai!

Echoes of festivals are being heard all around and Diwali is knocking on the doors. And there is such an occasion today that there are start-ups and lakhs of farmers of the country on the same platform in the same premises. Jai Jawan, Jai Kisan, Jai Vigyan and Jai Anusandhan! In a way, we can see a live form of this mantra in this ceremony.

Friends,

Today, all the important stakeholders of Bharat's farming sector are directly and virtually connected with us in this programme from every nook and corner of the country. Today, many important steps are being taken in the direction of making the life of the farmers easier, making the farmers more prosperous and making our farming systems more modern from such an important platform. Today, more than 600 Pradhan Mantri Kisan Samriddhi Kendras (PMKSKs) are being opened in the country. I have just been to an exhibition which has been organized here. There is so much innovation out there that I wanted to stay there a little more, but I did not want you to wait any further since it is the season of festivals, and therefore, I came to the podium. I really congratulate Mansukh Bhai and his team for creating such an innovative model in the form of PMKSK. It is not just a sale-purchase centre of fertilisers for the farmers, but this Kisan Samriddhi Kendra is a hub connecting with the farmers, responding to all their queries and helping them in every way.

Friends,

A short while ago, crores of farmers of the country got another installment of Rs 16,000 crore in the form of PM Kisan Samman Nidhi. Those who are sitting here can check their mobile phones and a message must have come by now that Rs 2,000 have been deposited in your account. There is no middleman and no company!

Interesting Facts

We stopped black marketing of urea by 100 per cent neem coating. We have worked hard to restart the country's 5 largest urea factories which were closed for years.

Do You Know?

How big the crisis was in the fertiliser sector before 2014, how black marketing of urea was prevalent, how farmers' rights were snatched away and in return farmers had to face lathicharge, our farmer brothers and sisters can never forget those days.

Money goes directly into the farmers' accounts. I congratulate all our beneficiary farmer families in every nook and corner of the country for receiving this money ahead of Diwali and important agricultural works.

I would also like to congratulate the agriculture start-ups taking part in this event for their new innovations for the betterment of the farmers, for lessening the burden of the farmers, for saving their money, for speeding up their work and for ensuring more yield in their small fields. Many such important things have been done by the youths who are involved in the start-ups. There are so many innovations visible here. I congratulate all such youths who are joining hands with the farmers and heartily welcome them to participate in this event.

Friends,

Today, a plan to provide affordable and quality fertilisers to farmers under the Bharat Brand in the form of One Nation, One Fertiliser, has started. Our farmer brothers and sisters can't forget the days before 2014 when there used to be a major crisis in the fertiliser sector in the form of the black marketing of urea. The rights of the farmers were snatched away and they had to face "lathis" (sticks). Major urea factories in the country were closed years ago. A new system was set up. Coffers of many people used to be filled through the import of urea, and therefore, they drew pleasure in closing down the factories in our country. We prevented black marketing of urea by mandating the production of 100 per cent neem coated urea. We have worked hard to revive the country's five largest urea factories, which were closed for years.

Friends,

Now Bharat is rapidly moving towards self-sufficiency in urea with revival of closed plants and through adoption of liquid Nano Urea. Nano Urea costs less and gives better yield. One bag of urea has now been replaced with a small bottle of Nano Urea. This is the achievement of science and technology. Now the farmers are getting rid of the problems of carrying bags of urea, the cost of transportation and also to keep a separate space in the house to store it. Along with other purchases, you can buy just a bottle of urea in the market and your work is done.

Today, two more major changes and reforms are being added to our efforts in the fertiliser sector. First, a campaign has been started from today to develop more than 2.5 lakh fertiliser shops across the country as PMKSKs. These will be such centres where not only fertilisers will be available, but also seeds, equipment, facilities for

Interesting Facts

A campaign is being started to develop more than 2.5 lakh fertiliser shops across the country as PMKSKs. These will be such centres where not only fertilisers will be available, but also seeds, equipment, soil testing and every kind of information & services that the farmer needs will be available at one place at these centres.

Do You Know?

Nano Urea is a means for more crop production at less cost. Just imagine, the work that requires one bag of urea can now be done by a small bottle of Nano Urea.

soil testing and other relevant information required by the farmers.

Secondly, our farmer brothers and sisters are getting rid of the problem of wandering from one place to another for fertiliser services. Until now, companies used to promote their brands of fertilisers through campaigns and offer maximum commission to those selling their brands of fertilisers. As a result, those getting maximum commission used to sell particular brands of fertilisers and would not sell other brands on which they get little commission. As a result, the farmers were under distress to get the quality fertilisers according to their needs due to competitions, different names and because of the arbitrariness of the agents selling fertilisers. The farmer was also confused if his neighbor bought a different brand. He would often dump the fertiliser that he had bought earlier and would buy the branded fertiliser which his neighbor had bought. Due to this confusion, sometimes the farmers used to spend double the amount on fertilisers.

Now suppose, if he desires a particular brand and if he does not get that particular brand, he would tend to use double the amount of fertiliser in his field for he is not too sure about the quality of the fertiliser that he bought. As a result, it would cost him more. All these problems have been addressed now.

Now the farmers are going to get rid of all kinds of confusion and better fertiliser will be available to them under the One Nation,

One Fertiliser scheme. Now, there will be only one brand for quality urea and that is "Bharat", which will be sold across the country. Now, urea will be available in the country only under the brand name of "Bharat". When there will be only one brand of fertiliser across the country, then the unfair competition over fertilisers of different companies will also end. This will also reduce the cost of fertiliser and it will be available in sufficient quantities easily.

Friends,

Today, about 85 per cent of our farmers in the country are small farmers. They do not have more than one hectare or one-and-a-half hectares of land. Moreover, with the passage of time, when a family expands, then such a small piece of land gets fragmented further into smaller pieces. And nowadays, we are also facing the problem of climate change. Diwali is almost here, but rain refuses to go away. Natural disasters continue to strike the country.

Friends,

Similarly, if the soil is inferior, if the health of our Mother Earth is not good, if our Mother Earth remains ill, then the fertility of our Mother Earth will also be affected. There will be more problems if the quality of water also gets worse. A farmer experiences all these

Interesting Facts

Today, 22 crore soil health cards have been given to the farmers in the country so that they can get correct information about the health of the soil. In the last 7-8 years, more than 1,700 varieties of such seeds have been made available to the farmers, which can fulfill their purpose and remain adaptable even in these changing climatic conditions.

Do You Know?

Now with "One Nation, One Fertiliser", farmers are going to get freedom from all kinds of confusion and better fertilisers are also going to be available. Now go to any corner of Bharat, urea will be sold with the same name, same brand and same quality—and this brand is Bharat!

problems in his everyday life. In such a situation, we have to create new systems in farming and adopt more scientific methods with an open mind to increase the agricultural yield.

Keeping this in mind, we have laid more emphasis on increasing the use of scientific methods in agriculture. Today, 22 crore soil health cards have been issued to the farmers in the country so that they get accurate information about the health of the soil. We are making conscious efforts in a scientific way to ensure that the best quality seeds are available to the farmers. In the last seven to eight years, more than 1,700 varieties of such seeds have been made available to the farmers which can withstand the changing climatic conditions.

In order to increase the quality of the seeds of the traditional coarse grains like millets, many hubs are being built in the country today. With the efforts of the government to encourage Bharat's coarse grains all over the world, next year has also been declared as the International Year of Coarse Cereals. Our coarse grains are going to be discussed all over the world. Now the opportunity lies before you how to conquer the world.

All of you are also well aware of the work that has been done regarding irrigation in the last eight years. There is a tendency of our farmers to fill their fields with water till the entire crop is submerged in water and turn the whole field into a pond. As a result, there is wastage of water, soil is also ruined and crops get destroyed. We have also worked in this direction by getting the farmers out of this situation. We are giving a lot of emphasis on per drop more crop, i.e., micro irrigation. We are emphasizing drip irrigation and sprinkler.

Earlier our sugarcane farmers were not ready to believe that sugarcane can be cultivated with less water. Now it has been proved that even with sprinklers, sugarcane cultivation can be very good and water can be saved. A farmer had this thing in his mind that just as if an animal is given more water, it will give more milk, and similarly, if a sugarcane field is given more water then more sugarcane juice will come out. Such has been the state of affairs in our country. About 70 lakh hectares of land in the country has been brought under the purview of micro-irrigation in the last seven-eight years.

Friends,

Natural farming also provides an important way to solve the challenges of the future. We are experiencing a lot of awareness across the country today in this regard. Farmers are working on a large scale in Gujarat, Himachal Pradesh and Andhra Pradesh

Interesting Facts

In the last seven to eight years, about 70 lakh hectares of land in the country has been brought under micro irrigation.

Do You Know?

If the health of our Mother Earth does not remain well, then her fertility will also decrease. If the health of the water deteriorates then there will be more problems. A farmer experiences all this in his everyday life.

as well as in Uttar Pradesh and Uttarakhand for natural farming. In Gujarat, plans are also being made at the district and gram panchayat level in this regard. The way natural farming has been promoted and got new markets in the past few years, the agriculture production has also increased manifold.

Friends,

The PM Kisan Samman Nidhi is also an example of how small farmers benefit from the use of modern technology. Since the launch of this scheme, more than Rs 2 lakh crore has been directly transferred to the bank accounts of farmers. This support proves handy when the farmers have to buy seeds or fertilisers. This is a huge expenditure for more than 85 per cent of the small farmers of the country. Today, farmers across the country tell me that the PM Kisan Samman Nidhi has taken care of their huge burdens.

Friends,

Today, we are also bridging the distance between the farm and the market by using better and modern technology. The biggest beneficiary of this is our small farmer, who is engaged in producing perishable products like fruits, vegetables, milk, fish, etc. Small farmers have also benefited a lot through Kisan Rails and Krishi Udan air service. These modern facilities are connecting farmers' fields to major cities across the country and markets abroad.

One of the results of these efforts is that agricultural goods are now being exported to those countries, which no one could have imagined earlier. Talking about agricultural exports, Bharat is one of the 10 major countries of the world. Our agricultural exports have grown by 18 per cent despite the constraints of Covid-19 in the last two years.

Dragon fruit, which is called "*Kamalam*" in Pahari language, is being exported on a large scale from Gujarat. Black garlic has been exported for the first time from Himachal. There are many fruits like Burmese grapes from Assam, apricots from Ladakh, bananas of Jalgaon or Bhagalpuri Zardalu mangoes that are appealing to foreign markets. Such products are being promoted today under schemes like One District One Product. Today, export hubs are also being set up at the district level, which is benefitting the farmers.

Friends,

Today our share in processed food is also increasing a lot. This is opening the way for farmers to get higher prices for their produce. The coarse grain of Uttarakhand was exported to Denmark for the first time. Similarly, organic jackfruit powder from Karnataka is also reaching new markets. Now, Tripura is also getting ready for this. We have sown these seeds in the last eight years, the fruit of which has started ripening now.

Interesting Facts

PM Kisan Samman Nidhi is also an example of how small farmers benefit from the use of modern technology. Since the launch of this scheme, more than Rs 2 lakh crore has been transferred directly into the bank accounts of farmers.

Do You Know?

Today, by using better and modern technology, we are also reducing the distance between farm and market. The biggest beneficiaries of this are our small farmers, who are associated with perishable products like fruits, vegetables, milk and fish.

Friends,

Let me share some statistics with you and you will realize how progress and change happens. While there were only two big food parks in the country eight years ago, today this number has gone up to 23. We are now making efforts to connect farmer producer organizations (FPOs) and self-help groups of sisters to this sector as much as possible. The government is making continuous efforts today to ensure involvement of small farmers directly in the cold storage, food processing and exports.

Friends,

The use of technology is resulting in big changes in the whole system, from providing seeds to ensuring markets. Our agricultural markets are also being modernized. At the same time, e-NAM (National Agriculture Market) has been set up so that farmers can use the technology to sell their produce in any market of the country while sitting at home. So far, more than 2 crore farmers and 2.5 lakh traders of the country have joined e-NAM.

You will also be happy to know that transactions worth more than Rs 2 lakh crore have been done through e-NAM so far. You must have noticed that property cards are being given to the farmers by mapping their lands and houses. Modern technology like drones is being employed for all these efforts.

Friends,

Our start-ups can take the use of modern technology to a new era to make agriculture more profitable. Today, a large number of start-up partners are present in this event. Take note of this figure. Earlier, there were only 100 start-ups working in the agriculture sector. In the last seven-eight years, this figure has jumped to more than 3,000. These start-ups, these innovative youths and these talents of Bharat are rewriting the future of Bharat's rural economy. Our start-ups have the solution to every problem, from production to distribution.

Kisan Drones are going to make life much easier for the farmers. Drones can predict the fertiliser needed for the soil, how much irrigation is needed and which pesticide is needed. The drones can give you the right guidance. The drone sprays the pesticide where it is required. It will also prevent the wastage of fertiliser, and my farmer brothers and sisters will also be saved from the ill-effects of chemicals.

Brothers and sisters,

Today there is another big challenge which I would like to mention before all our farmer friends and our innovators. We all need to work in mission mode by understanding why I am giving so much emphasis on self-reliance in agriculture and what is the role of farmers in this regard. Today, we spend the most on importing

> Interesting Facts
>
> **Small farmers have also got a lot of benefits from Kisan Rail and Krishi Udaan air service. Today, these modern facilities are connecting farmers' fields with big cities across the country and foreign markets.**

Do You Know?

When we talk about agricultural exports, Bharat is among the top 10 countries of the world. Despite the disruptions of Covid-19, despite two difficult years, our agricultural exports have increased by 18 per cent.

edible oil, fertilisers and crude oil. Every year lakhs of crores of rupees are paid to other countries just to buy them. When there is a problem abroad, it has an effect on us too.

Covid-19 did not go from our lives completely, and simultaneously war broke out in the region from where we used to import so many goods. The countries which used to meet most of our requirements are embroiled in war. The impact of the war on such countries has also increased.

For example, fertiliser! Be it urea, diammonium phosphate (DAP) or other fertilisers, they have become so expensive in the world markets and our country is bearing so much economic burden. Today, we buy urea from abroad at Rs 75-80 per kg. In order to ensure that the farmers of our country are not burdened and they do not face any crisis, we are providing urea to them at Rs 5-6 per kg, which we are importing at Rs 75-80 per kg. In order to ensure that farmers get fertilisers at a low price, it takes a toll on the government exchequer. We have to spend about Rs 2.5 lakh crore only on the purchase of fertilisers this year.

Brothers and sisters,

To reduce the expenditure on imports, to make the country self-reliant, we all have to take a common resolution and walk in that direction together. We also have to make a resolve against importing food and agricultural products. A lot of work is going on in the country today on biofuel and ethanol to reduce the foreign dependence on crude oil and gas. The farmers and our farming communities are directly connected with this. Work is in progress so that cars can run on ethanol produced from farmer's produce, and bio-CNG and biogas are produced from waste and cow dung. We have also started Mission Oil Palm for self-sufficiency in edible oil.

Today I would urge all the farmer friends to take maximum advantage of this mission. We can greatly reduce the import of edible oil by increasing the production of oilseeds. The farmers of the country are fully capable in this regard. When I called you to work on mission mode on pulses in 2015, you took my words seriously and proved it.

Otherwise, what was the situation earlier? We had to import pulses from abroad for our consumption. When our farmers decided, they increased the production of pulses by almost 70 per cent. With such will power, we have to move forward and make Bharat's agriculture more modern and take it to new heights. With the resolve that we will make agriculture attractive and prosperous in the "Amrit Kaal" of independence, I convey my best wishes to all my farmer brothers and sisters and to all the youths associated with start-ups.

Thanks a lot!

(DISCLAIMER: This is an approximate translation of PM's speech. Original speech was delivered in Hindi.)

❑ ❑ ❑

Preface

Bharat is known as the country of villages. When anyone calls Bharat a country of villages, it naturally conveys the message that Bharat is a country of farmers. For a long time, Bharat has been recognized throughout the world as an agricultural country. How important the role of farmers and farming is from the Bharat's perspective can be estimated from the fact that almost two-thirds of the country's population is dependent on agriculture in some form or the other. We have many industries that depend on agriculture for raw materials. In such a situation, Prime Minister of Bharat Shri Narendra Modi ji has been continuously working since the beginning of his tenure to ensure that Bharat's agricultural sector continues to progress and that it does not face any crisis.

The Prime Minister has often said that his government is a government of farmers and the poor. He has not only said this but has proved it time and again through the policies and decisions of his government. Whenever any issue related to farmers has come up, the Prime Minister has taken personal interest in it and has taken a decision always in favour of the farmers. There were many occasions when big countries of the world chose the path of putting the economic burden on their farmers under the pressure of global situations, but Modi ji remained committed that we should not increase the economic burden on our farmers.

In the last few years, many countries of the world faced the problem of fertiliser crisis. But under the leadership of the Prime Minister, Bharat did not let these crises affect the farmers and agriculturists of our country. Modi ji ensured that the rising prices of fertilisers in the global markets would not affect the farmers of the country. For this, the government had to increase the subsidy on multiple occasions. Even though the burden of subsidy on the government continued to increase, the Prime Minister was determined not to let the country's farmers feel its burden.

Interesting Facts

Today, the Prime Minister's vision of "self-reliant Bharat" (Aatmanirbhar Bharat) is being realized in many areas. Fertiliser sector is also one of them. During the tenure of Modi ji, the fertiliser sector has succeeded in writing a golden story of development. Today, Bharat is moving rapidly on the path of becoming self-reliant in fertilisers.

Do You Know?

Modi ji ensured that the rising prices of fertilisers in the global markets should not affect the farmers of the country. For this, the government increased the subsidy. Even though the burden of subsidy on the government continued to increase, the Prime Minister was determined not to let the country's food providers feel its burden.

When fertiliser prices were increasing all over the world, a fertiliser supply chain disruption erupted in many countries. But on this front also, under the guidance of the Prime Minister, the Government of Bharat ensured that there is no shortage of fertilisers in the country. Bharat worked to ensure uninterrupted supply of fertilisers to its farmers at reasonable prices by entering into long-term agreements with countries that manufacture fertilisers and supply raw materials for fertilisers.

When Covid-19 hit the whole world, in order to deal with that challenge, Prime Minister Shri Narendra Modi ji gave the mantra of "self-reliant Bharat" (Aatmanirbhar Bharat) to the entire nation. But if anyone has seriously observed Modi ji's tenure as Prime Minister from the first day, i.e., May 26, 2014, then that person will know that even though the Prime Minister had put forward the vision of "self-reliant Bharat" in front of the country in 2020, he started working on it as soon as he became the Prime Minister. When he talked about "Make in India" soon after he became the Prime Minister, it was clear that he wanted to move Bharat forward in the direction of making it self-reliant. The imprint of his efforts to make Bharat self-reliant is clearly visible in many areas.

During the tenure of Prime Minister Shri Narendra Modi ji, the country has set many new records of progress. Today, the Prime Minister's vision of "self-reliant Bharat" is being realized in many areas. Fertiliser sector is also one of them. During the tenure of Modi ji, the fertiliser sector has succeeded in writing a golden story of growth. Today, Bharat is moving rapidly on the path of becoming self-reliant in terms of fertilisers. Approximately 350 lakh tonnes of urea is consumed every year in Bharat. Of these, Bharat imports approximately 70-80 lakh tonnes. Now, it is continuously decreasing and we hope that soon Bharat will become self-reliant in urea. With the inspiration of the Prime Minister, the closed urea factories were made operational again. Besides this, new urea production plants were also set up. Apart from this, by developing Nano Urea and producing it on a large scale, Bharat has made significant progress in becoming self-sufficient in urea. Today, farmers of the country are adopting Nano Urea on a large scale. In the case of Nano DAP as well, remarkable work is being done towards making the country self-reliant.

Bharat became the first country in the world to successfully start commercial production of the revolutionary Nano Urea.

Interesting Facts

Approximately 350 lakh tonnes of urea is consumed every year in Bharat. Of these, Bharat has been importing approximately 70–80 lakh tonnes. Now it is continuously decreasing and we hope that soon Bharat will become self-sufficient in urea.

Do You Know?

With the inspiration of the Prime Minister, the closed urea factories were being made operational again. Besides this, new urea production plants were also set up. Apart from this, by developing Nano Urea and producing it on a large scale, Bharat has made significant progress in becoming self-sufficient in urea.

Similarly, Bharat also created history by developing Nano DAP. Farmers are getting many benefits from the use of Nano Urea and Nano DAP. They are cheaper than conventional urea and conventional DAP in terms of prices. Due to this, farmers are benefiting financially. Instead of a urea bag of around 45 kg, a half-litre bottle is being used. This is providing convenience to the farmers in the use of fertilisers. This topic has been explained in detail in this book.

The Prime Minister often talks about balanced use of fertilisers. He also said that increasing the use of organic fertilisers instead of chemical fertilisers will be a great service to Mother Earth. Taking inspiration from this, the Central Government has come up with a scheme like PM-PRANAM to promote balanced use of chemical fertilisers. Under this, there is a provision to give incentives to farmers and states who make balanced use of fertilisers. Also, GOBARdhan (Galvanizing Organic Bio-Agro Resources Dhan) Scheme is being run to promote organic fertilisers. The Central Government is working at many levels to ensure that alternative fertilisers are available on a large scale in the country and farmers adopt them widely. Detailed information about these works is given in this book. Decisive and result-oriented steps were taken against black marketing and hoarding of fertilisers in the country.

Along with many such successes, the story of how the country is moving forward on the path of becoming "self-reliant Bharat" (Aatmanirbhar Bharat) in the field of fertilisers, has been presented in an interesting manner through this book.

– Dr. Mansukh Mandaviya

❑ ❑ ❑

"Big urea factories in the country had been closed years ago. Because a new world order had come up, imports used to fill many homes and pockets; hence they were happy with the closure of the factories here. We have worked hard to restart the country's five largest urea factories which were closed for years."

"To reduce the expenditure on imports, to make the country self-reliant, we all have to take a common resolution, we all have to move together in that direction."

1

Ensuring Fertiliser Security

I have been associated with farming since childhood. I was born in a farming family in Hanol village near Palitana, Bhavnagar, Gujarat. My father was a farmer. Therefore, working in farms started since childhood. Even at home, various aspects of farming were discussed and I used to see its practical application in the fields. This marked the beginning of my childhood training in the basics of agriculture. As I grew older, I started helping my father and brothers in the fields. From a young age, I have seen work ranging from ploughing to managing irrigation and spraying fertilisers.

I hugely benefitted from these experiences. Firstly, I got a chance to know and understand various aspects related to agriculture. The second advantage was that I was able to know where farmers face problems in farming. The real understanding about the problems of farmers developed gradually. From this, I came to know what work should be done if the agriculture scenario is to be changed at the grassroots level.

Even after joining politics, my association with farming continued. After becoming an MLA, when Shri Narendra Modi ji was the Chief Minister of Gujarat, he gave me the responsibility of the Chairman of Gujarat Agro Industries Corporation Limited. While working in this organization, my experience of working in the fields since childhood came in handy. But an even bigger role was played by Modi ji's teachings at every step of my journey. Modi ji taught me that if I want to understand any problem, then I should go down to the grassroot level, sit with the last-mile person, talk to him, understand his experiences and then we will find an effective way to solve those problems. Modi ji has his special working style as an administrator. He prefers to listen directly to the concerned parties. If there was any problem in any district, he used to take presentations directly from the Collector. He used to spend a lot of time, listen to each and every thing carefully and then find a solution to it. Taking inspiration from this working style of Modi ji, I also tried to work in a similar manner as Chairman of Gujarat Agro Industries Corporation Limited and was able to make my small contribution in the overall development of agriculture in Gujarat.

I would like to mention one work here. Many types of fruits are exported from Gujarat on a large scale. But the process of exporting fruits takes many days to reach the target market or country. To ensure that the fruits do not spoil during the transit period, it was necessary to do irradiation of the fruits. Till that time, there was no government irradiation centre in Gujarat. The farmers of Gujarat had to load their fruits in trucks and take them to Mumbai. Trucks

> Interesting Facts
>
> **Modi ji taught us that if we want to understand any problem, then we should go down to the grassroots level, sit with the last-mile person, talk to him, understand his experiences and then we will find an effective way to solve those problems.**

Do You Know?

There are three major factors of agricultural productivity. The first of these is improved varieties, the second is irrigation and the third is fertiliser.

loaded with fruits from Gujarat used to stand in queue for hours in Mumbai waiting for their turn. In this process, the fruits of many farmers used to get spoiled and they used to suffer huge losses.

In such a situation, on the instructions of Modi ji, we decided that we have to develop an irradiation centre in Gujarat, thus saving the time of farmers who were required to go to Maharashtra for this work. Due to this, farmers' expenses on freight transportation started reducing and wastage of fruits was also avoided. Their fruits spoiled less and hence were safe during transit. Because of this, their export earnings increased and the prosperity of the farmers of Gujarat increased too. Also, the farmers of Gujarat were able to increase their income by selling their produce in larger quantities in foreign markets by supplying quality fruits at more competitive prices than before. Modi ji's leadership and inspiration along with on-ground knowledge and experience of agriculture helped me in working effectively as Chairman of Gujarat Agro Industries Corporation Limited. After being the Chairman, Shri Narendra Modi ji made me a Rajya Sabha MP. During that time, I raised many questions related to farming in the Parliament. Wherever I have been and whatever position I have held, I have tried to contribute for betterment of farmers and farming sector.

In this background, when on July 5, 2016, Prime Minister Shri Narendra Modi ji assigned me the responsibility to be in his council of ministers as the Minister of State in the Union Ministry of Chemicals and Fertilisers, it was a big opportunity for me to contribute something more for the farmers' welfare and farming. Modi ji guided me while I was in Gujarat and his working style inspired me to do better work, I decided to replicate the same in the Central Government, where I got the opportunity to serve farmers under the able guidance of Modi ji.

As a Union Minister of State in the Ministry of Chemicals and Fertilisers, when I started attending the meetings of the Department of Fertilisers, I understood that there are three major factors of agricultural productivity. The first is advanced varieties of seeds, the second is irrigation and the third is fertiliser. Of these, a lot of work was done on seed varieties and irrigation during the Green Revolution, but the persistent problems in the fertiliser segment could not be solved at the pace at which it should have been. Understanding these circumstances, I started working in the ministry. During this period, I constantly tried to understand the problems related to fertilisers from many Members of Parliament, chief ministers of states, ministers of states and small and big farmers from my village, state and country. I also had regular discussions with scientists working in the fertiliser sector. Apart

Interesting Facts

Modi ji has his special working style as an administrator. He used to communicate directly to the concerned parties. If there was any problem in any district, he used to take presentations directly from the Collector. He used to spend a lot of time, listen to each and every thing carefully and then find a solution to it.

Do You Know?

Many old urea factories in Bharat were closed. The import of urea from China and other countries was continuously increasing. Bharat used to import about one-third of its urea requirements from other countries.

from high level meetings with senior officers in the Ministry, I tried to understand the fertiliser sector by sitting for hours even with officers down the line to the level of Section Officers.

During this entire process, I noticed many things like Bharat is at second place in the world in terms of the production of fertilisers and our country is also at second place in terms of their imports. It was also revealed that Bharat is included in the three major agriculture-based countries of the world. In such a situation, if the country has to move forward, then the agriculture sector will have to progress continuously. The progress of the agricultural sector largely depends on the continuous supply of quality and affordable fertilisers. It was also understood that there is a need to increase the level of awareness regarding the proper use of fertilisers by the farmers in the country at the grassroot level. But even before that, a need was felt to ensure the uninterrupted supply of quality fertilisers at reasonable prices in the states and the country.

Many old urea factories in Bharat were closed. The import of urea from China and other countries was continuously increasing. Bharat used to import about one-third of its urea requirements from other countries. This meant that our dependence on imports was high to ensure the country's food security. Under these circumstances, Prime Minister Shri Narendra Modi ji has entrusted us with the responsibility of improving the systems in such a way that adequate availability of fertilisers at the reasonable price can be ensured up to the last village and the last farmer of the country. He also instructed us that we have to move rapidly towards self-reliance (Aatmanirbhar) in the fertiliser sector. Modi ji also guided us to shift towards alternative fertilisers.

Based on his guidance, we started working in the ministry. From my past experiences of living in the village and being a farmer myself, I understood the path to alternative fertilisers. In the village, we used to use livestock dung as fertiliser in agriculture. Therefore, taking advantage of those experiences, we started working at the policy level. We had to fulfill Prime Minister Shri Narendra Modi's resolve of "self-reliant Bharat" (Aatmanirbhar Bharat) in every sector. With this view, the Ministry started working to achieve self-reliance in fertilisers and minerals also. Countries like Saudi Arabia, Morocco, Oman, China and Jordan have dominance in fertiliser minerals, as fertilisers and minerals are available in abundance as raw materials in these countries. We started working in this direction to study and start mining the fertiliser minerals wherever they are likely to be available in the country. We also achieved success in this direction. Our government continuously worked to increase the availability of raw materials in the country. As Modi ji says, we have to work in coordination

> **Interesting Facts**
>
> **On the directions of Modi ji, we decided that we have to develop an irradiation centre in Gujarat itself. So that the farmers of Gujarat do not have to go to Maharashtra for this work and they can export their fruits without any loss.**

Do You Know?

Countries like Saudi Arabia, Morocco, Oman, China and Jordan enjoy dominance in fertiliser minerals. Because fertilisers minerals are available in abundance as raw materials in these countries. We started working in this direction to study and start mining the fertiliser minerals wherever they are likely to be available in the country. We also got success in this direction.

between ministries, and we got support from the Ministry of Mines, which identified fertiliser mineral reserves which could be used in the production of fertilisers in states like Rajasthan and Madhya Pradesh. In these states, mining of rock phosphate was increased with the help of Geological Survey of India (GSI). Efforts to increase the production of indigenous SSP and DAP were made by the Ministry of Chemicals and Fertilisers with the help of all stakeholders. This has been discussed in more detail in a separate chapter of this book.

Considering our import dependence, fertilisers were sold at higher rates in the international market. But under the leadership of Prime Minister Shri Narendra Modi ji, we also succeeded in changing this situation in our favour. We managed to convince the major fertiliser suppliers across the world that if we are the second largest importer, then they need us as much as we need them, hence we should also have an important say in determining the prices. Positive results emerged because of these continued efforts.

In July 2022, one ton of DAP was being brought to Bharat at a cost of $945. The war in Ukraine and Russia caused prices to rise and the entire supply chain was disrupted. But in July 2023 the cost came down to $440. Phosphoric acid is the most important component of making phosphatic fertilisers. During the same period mentioned earlier, phosphoric acid fell from $1,715 per tonne to $970 per tonne. Rock phosphate was at $305 per tonne in October 2022. By July 2023, it came down to $208 per tonne. Sulphur has special importance in the fertiliser sector. Sulphur is the important raw material for manufacturing sulphuric acid. Sulphur was at $279 per tonne in July 2022. By July 2023, prices of sulphur came down to $90 per tonne. Ammonia, an important raw material, was at $1,100 per tonne in August 2022 which dropped to $285 in July 2023.

The question that naturally arose was how did all this happen? The answer to this is that Bharat took forward the diplomatic- and ministerial-level talks in a very proactive manner. Our companies entered into long-term supply agreements with the companies of those countries from where we source fertilisers or raw materials for the production of fertilisers. Besides long-term agreements, companies also went for establishing joint ventures. This resulted in a decline in prices and uninterrupted supply of fertilisers to farmers across the country at reasonable prices.

Interesting Facts

Bharat is included in the three major agriculture-based countries of the world. In such a situation, if the country has to move forward, then the agriculture sector will have to progress continuously. The progress of this sector largely depends on the continuous supply of quality and affordable fertilisers.

Do You Know?

We managed to convince the major fertiliser suppliers across the world that if we are the second largest importer, then they need us as much as we need them. Hence, we should also have an important say in determining the prices. Our efforts gave positive results.

Inspired by this vision of fertiliser security, we started to work on many levels to break the monopoly of some fertiliser producers and fertiliser-producing countries in the international market. On one hand, we revived the closed urea factories, while also establishing new plants. We also took steps to promote the production and use of alternative fertilisers. Bharat became the first country in the world to start the commercial production of Nano Urea and Nano DAP.

My experience of working as Shipping Minister in the Government of Bharat proved handy in this direction. The movement of fertilisers and related raw materials in international waters, their origin points and economics behind all this helped me in making necessary interventions to secure the country's interest.

New technology not only brings transparency in any field but also effectively gives us information about all aspects. In fertiliser sector, we started Integrated Fertiliser Management System (iFMS). It started capturing end-to-end data related to production, import, availability, distribution and shortfalls. This increased the efficiency of the entire supply chain of fertilisers. On September 1, 2017, we started Direct Benefit Transfer (DBT) in the fertiliser sector through Point of Sale Machine (POS), in which subsidy was being transferred to fertiliser companies on the basis of sales of subsidised fertilisers to the farmers. After the success of the first phase, it was implemented in the entire country by March 2018. Today, subsidies are being transferred directly on the basis of the data received from this system.

A dashboard has been developed under iFMS. This system provides accurate information regarding the cost of fertiliser manufacturing and import price of fertilisers. This ensures that companies cannot sell fertiliser at a rate higher than the prescribed rate. It is not possible for companies to sell fertilisers at arbitrary rates to increase their profits. The estimate of fertiliser consumption for the entire season given by the Ministry of Agriculture is also visible on the dashboard and every important information about the supply chain of fertiliser is available here in a completely transparent manner.

Through this dashboard, massive data is generated on various dimensions related to the supply and consumption of fertilisers. Based on captured data, "evidence-based policy making" has increased in the fertiliser sector and effective decisions are being taken.

Interesting Facts

Prime Minister Shri Narendra Modi ji assigned the responsibility to improve the systems in such a way that adequate availability of fertilisers at the right price can be ensured to the last village and the last farmer in the country. Modi ji asked us to move towards alternative fertilisers.

Do You Know?

Inspired by the vision of fertiliser security, we started work on many levels to break the monopoly of some fertiliser producers and fertiliser-producing countries in the international market. On the one hand, we have revived the closed urea factories and on the other hand, we also started the work of starting new plants.

We held frequent meetings with agriculture ministers of states and had regular dialogues with chief ministers and officials, understanding their problems in detail. We sensitized them about the dashboard and shared the effective strategy of fertiliser management using the iFMS dashboard.

Prime Minister Shri Narendra Modi ji always taught us that "national interest" and "farmer interest" are of paramount importance. In the fertiliser sector, we continuously worked with this basic mantra. Since Prime Minister Shri Narendra Modi's government thinks in "total" and not in "token", we worked in same spirit to achieve the desired results in the fertiliser sector.

We followed a true team spirit by maintaining regular communication with all stakeholders like industry, state agriculture departments, Union Ministry of Agriculture and Ministry of Railways to ensure uninterrupted supply of fertilisers. Here, I would like to specially mention the role of Railways executed through Director (Movement), for extraordinary domain knowledge and its execution in true spirit during these years. The number of challenges we faced and unparalleled responses made

Do You Know?

Work was also done to promote the production and use of alternative fertilisers. Bharat became the first country in the world to start commercial production of Nano Urea. We also worked to promote the production and use of organic fertilisers.

on his position are worth appreciation. I will always be indebted to all this.

We got great support from various ministries like Port, Shipping and Waterways when it comes to movement of fertiliser-carrying ships; Ministry of External Affairs to make necessary diplomatic interventions in source countries; Ministry of Finance to meet the financial needs in terms of subsidy; Ministry of Petroleum for supply of gas; and many more.

The Prime Minister guided us at every step in all these works. It is the result of the efforts made under the leadership and regular guidance of Prime Minister Shri Narendra Modi ji that the country is moving rapidly on the path of fertiliser self-sufficiency while ensuring "fertiliser security".

❑ ❑ ❑

"Why am I giving so much emphasis on Aatmanirbharta and what is the role of farming and farmers in this? We all need to understand this and work in mission mode. Today, we spend the most on importing edible oil, fertilisers and crude oil. Every year we have to pay lakhs of crores of rupees to other countries just to buy these goods. When any problem arises in foreign countries, its impact falls on us here also."

"Now take the example of fertiliser. Be it urea, DAP or other fertilisers, these are becoming expensive in the world markets, causing so much economic burden that our country has to bear. Today we buy urea from abroad at the rate of Rs 75-80 per kg. But the farmers of our country should not be burdened, our farmers should not face any new crisis. Today, we bring urea from outside for Rs 75-80 per kg, we deliver it to the farmers at Rs 5-6 per kg, so that my farmers brothers and sisters should not suffer."

2

Global Fertiliser Scenario

Bharat ranks second in the world in terms of production of fertilisers. But along with this, Bharat is also the second largest importer of fertilisers in the world. Bharat ranks second after China in the production of nitrogenous fertilisers and phosphate fertilisers. Bharat also ranks second in the world in terms of consumption of these two fertilisers. Obviously, in such a situation, the conditions of the international fertiliser market become very important for us. Although we have made considerable progress in the field of fertilisers in the last few years, the scenario at the international level has been quite volatile, which posed various challenges.

Due to various geopolitical developments, the supply of fertilisers in the international market was affected and prices also witnessed a sharp rise. Despite this, our government has ensured that farmers do not face any problems regarding the supply and pricing of fertilisers.

Highlighting these efforts, while interacting with farmers in Sikar, Rajasthan on July 27, 2023, Prime Minister Shri Narendra Modi ji said, "I want to tell this truth about fertiliser prices to every farmer brother and sister of the country. Today, the bag of urea that we give to the farmers in Bharat for Rs 266, the same bag of urea is available to the farmers in our neighborhood in

Pakistan for around Rs 800. Today, the bag of urea that we give to farmers in Bharat for Rs 266, the same bag of urea is available to the farmers of Bangladesh market for Rs 720. Today, the bag of urea that we give to farmers in Bharat for Rs 266 is the same bag of urea that farmers in China get for Rs 2,100. And do you know, how much it costs for the same bag of urea in USA these days? American farmers have to spend more than three thousand rupees for the same bag of urea for which you pay less than Rs 300. Can you imagine the difference between three hundred and three thousand? Our government will not let the farmers of Bharat face problems due to the prevailing international prices of urea. Our policies stand testimony to the truth that the farmers of the country are witnessing and experiencing every day. When he goes to buy urea, he is sure that it is Modi's guarantee. If you ask the farmer what is called guarantee, he will know it."

This shows how effectively Bharat dealt with the challenges of the global fertiliser scenario. The recent past has been a turbulent period for the international economy. While there was economic slowdown in some countries of the world, there was an atmosphere of war and unrest in some areas. No country in the world remained untouched by Covid-19 pandemic and it had a negative impact on every economy of the world. Under these circumstances, the international scenario of fertilisers was also affected, posing challenges to food security globally.

Interesting Facts

Bharat ranks second in the world in fertiliser production. But along with this, Bharat is also the second largest importer of fertiliser in the world.

Post Covid-19 pandemic, the fertiliser world was surrounded by new problems that had their genesis in geopolitics like Russia–Ukraine crisis, Hamas–Israel issue and their ripple effects all across.

In this era of globalization, whenever a situation of instability arises, it affects the economies of the entire world. But even amidst this environment of instability, our government under the leadership of Prime Minister Shri Narendra Modi ji ensured that these instabilities do not affect the farmers of Bharat and our agriculture continues to move forward on the path of progress.

If we look at the global scenario of fertilisers, it becomes clear that Bharat is dependent on imports in the case of phosphoric acid which is used as raw material in the production of many fertilisers. Data till 2020 shows that around 47 million tonnes of phosphoric acid is produced every year all over the world. China's share in this is about 37 per cent, whereas Morocco is at second place with a share of about 15 per cent. USA produces around 14 per cent of the total global production and it holds third position. Russia produces 8 per cent of the total global production of phosphoric acid. Almost 90 per cent of the total phosphoric acid production is used in the production of fertilisers in the same country.

Globally, approximately 40 lakh tonnes of phosphoric acid is exported to other countries. Morocco ranks first with a share of 43 per cent in total exports. Jordan is at second place with 15 per cent

Interesting Facts

Bharat ranks second after China in the production of nitrogenous fertilisers and phosphate fertilisers.

share and Senegal is at third place with 12 per cent share. China and Tunisia jointly rank fourth with exports of 8 per cent. More than half of the total global export, i.e., 23 lakh tonnes of phosphoric acid comes to Bharat. This means that Bharat's import dependence for the phosphoric acid is very high. To meet the fertiliser needs of Bharat, approximately 43 lakh tonnes of phosphoric acid is required every year. We import almost half of this. We have also set up joint ventures in Senegal, Jordan, Morocco, Tunisia and South Africa to meet its phosphoric acid requirements.

Similarly, if we look at the international scenario of DAP/MAP, it is found that approximately 65 million tonnes of it is produced every year in the whole world. Of this, a maximum of around 40 per cent production takes place in China. Morocco is at second place with about 12 per cent share in global production and Bharat is at third place with approximately 12 per cent share. The global trade of DAP/MAP is around 31 million tonnes worldwide. China has the highest share in this with about 28 per cent. Morocco is at second place with approximately 23 per cent share and Saudi Arabia is at third place with around 14 per cent share. USA is at fourth place with a share of about 13 per cent.

Bharat imports maximum DAP/NPK. Bharat imports approximately 6 million tonnes of DAP/NPK every year. Brazil is the second largest importer with imports of 5 million tonnes.

Interesting Facts

Bharat also ranks second in the world in terms of consumption of nitrogen and phosphate fertilisers, whereas Bharat ranks fourth in terms of consumption of potash.

Do You Know?

Production of Nano DAP has started in Bharat. Also, the use of SSP is being promoted so that the dependence of farmers on DAP is reduced. Obviously, when farmers' dependence on DAP will reduce, Bharat's import dependence on this fertiliser will also come down.

Bharat is second only to China in terms of usage of DAP/NPK. While China consumes 20 million tonnes of DAP/NPK every year, Bharatiya farmers use 11 million tonnes of DAP/NPK in their fields every year. Bharat imports more than half of its DAP/NPK requirements. To change this situation, production of Nano DAP has started in Bharat. Also, the use of SSP is being promoted so that the dependence of farmers on DAP is reduced. Obviously, when farmers' dependence on DAP will reduce, Bharat's import dependence on this fertiliser will reduce. Taking lessons from the international scenario of every fertiliser, Bharat is working effectively at various levels to ensure "fertiliser security". These steps have been discussed in detail in different chapters of this book.

❑ ❑ ❑

"Bharat is experiencing swift and significant transformations. Presently, there is a noticeable increase in people's confidence, trust in the government and their dedication to actively participate in shaping a new Bharat, which is apparent across various sectors."

“

Our work culture involves thinking big, thinking far. *Sarvajan Hitay, Sarvajan Sukhay* is central to our work ethos. This energy drives us to achieve more than what we think and aspire.

”

3

Quest in Urea Self-reliance

The Central Government, led by Prime Minister Shri Narendra Modi ji, has taken several steps to make the country self-reliant in urea. On one hand, Bharat became the first country in the world to commercially produce Nano Urea, while on the other hand, the Modi government also started new production units of conventional urea.

The efforts to increase investment in urea sector in Bharat started with new investment policy. The Central Government took an important step towards encouraging new investment in urea sector by amending the new investment policy on October 7, 2014. Under the new investment policy, the Modi government has started many urea production units across the country.

Under the new investment policy, Matix Fertilisers and Chemicals Limited have also started a new urea production unit in Panagarh, West Bengal. Its annual production capacity is 12.7 lakh tonnes. Chambal Fertilisers and Chemicals Limited has also started a new urea production unit at Gadepan, Rajasthan.

Apart from these, the Central Government has taken up the task of restarting four previously closed urea production units. Under the new investment policy, these four units where the production restarted are Ramagundam in Telangana, Gorakhpur in Uttar Pradesh, Sindri in Jharkhand and Barauni in Bihar.

Ramagundam Fertilisers and Chemicals Limited (RFCL) has started commercial production at its Ramagundam unit from March 22, 2021. To start urea production at Ramagundam, the government formed a joint venture with three public sector companies National Fertilisers Limited (NFL), Engineers India Limited (EIL) and Fertiliser Corporation of India Limited (FCIL). This joint venture was named Ramagundam Fertilisers and Chemicals Limited. In this, the share of both NFL and EIL was kept at 26-26 per cent. FCIL has 11 per cent stake in it. The annual production capacity of this factory is 12.7 lakh tonnes. Its foundation stone was laid by Prime Minister Shri Narendra Modi ji on August 7, 2016. More than Rs 6,000 crore were spent in starting this urea factory.

Hindustan Urvarak and Rasayan Limited has restarted the closed urea factory in Gorakhpur and started commercial production of urea from March 26, 2022. In fact, to restart the closed urea factories of Gorakhpur, Sindri and Barauni, the Central Government has directed National Thermal Power Corporation, Indian Oil Corporation, Coal India Limited, FCIL and Hindustan Fertiliser Corporation Limited (HFCL) to form a joint venture, Hindustan Urvarak and Rasayan Limited. In this, the share of the first three PSUs is 29.67 per cent and 11 per cent is of FCIL/HFCL.

Interesting Facts

The Central Government took an important step towards encouraging new investment in urea production units by amending the new investment policy on October 7, 2014. Under the new investment policy, the Modi government has started many urea production units across the country.

Do You Know?

The Central Government has taken up the task of restarting four closed urea production units. Under the new investment policy, these four units where the production restarted are Ramagundam in Telangana, Gorakhpur in Uttar Pradesh, Sindri in Jharkhand and Barauni in Bihar.

Each of these three urea factories has an annual production capacity of 12.7 lakh tonnes. The Prime Minister laid the foundation stone of the Gorakhpur factory on July 22, 2016 and the foundation stone of the Sindri factory on May 25, 2018. The foundation stone of Barauni Urea Factory was laid by the Prime Minister on February 17, 2019.

Similarly, the closed urea factory of FCIL in Talcher, Odisha is also being revived. This unit will produce urea through gasification of coal. The special subsidy policy for this unit was approved by the Central Government on April 28, 2021. The location where this unit is located is strategic. Coal mines are near to it. Therefore, the Fertiliser Department of the Central Government chose the route of coal gasification to restart this urea production unit. For this work, the Fertiliser Department formed a consortium of three companies: Rashtriya Chemicals and Fertilisers, Gas Authority of India Limited and Coal India Limited. This is the first urea production unit of its kind in the country. Therefore, the Central Government took the decision to have a special subsidy policy in place to ensure the execution of the project. This special policy will remain in force for eight years.

In fact, restarting the closed urea factories has its own background. Fertiliser Corporation of India Limited had five urea factories. These production units were in Sindri in Jharkhand, Talcher in Odisha, Ramagundam in Telangana, Gorakhpur in Uttar Pradesh and Korba in Chhattisgarh. Whereas three urea factories of HFCL were working in Barauni in Bihar, and Haldia and Durgapur in West Bengal. Both these companies were continuously incurring losses due to various reasons. Because of this, the Board for Industrial and Financial Restructuring declared both of them sick in 1992. In 2002, urea production from these factories stopped and all their employees were given voluntary retirement.

Since 2007, efforts have been made at different levels to explore the possibility of restarting these urea factories. But the turning point came in 2016 when the Union Cabinet, headed by Prime Minister Shri Narendra Modi ji, approved the decision to restart three urea production units on July 13, 2016. The Cabinet decided that the urea factories of Gorakhpur, Sindri and Barauni will be restarted. For this, it was also decided that a special purpose vehicle of public sector undertakings, i.e., a company would be formed. In this, it was decided to form a special company with the partnership of National Thermal Power Corporation, Coal India Limited and

Interesting Facts

With the restarting of the five old fertiliser plants, Bharat's annual urea production capacity will increase by about 63 lakh tonnes. With this, farmers across the country will continue to have adequate quantity of urea available for their crops.

Do You Know?

The commencement of production from these urea factories will help in bridging the gap between demand and supply of urea. Besides, small and medium enterprises will also develop around these factories and will make various supplies to these factories. This will create an environment of entrepreneurship at the local level and strengthen the local economy.

Indian Oil Corporation Limited with FCIL/HFCL. These companies were formed in such a way that the stake of FCIL/HFCL was at least 11 per cent and the stake of other public sector companies was at least 51 per cent.

Resumption of closed fertiliser factories has also created new employment opportunities. The fertiliser factory at Ramagundam is providing direct employment to 500 people. At the same time, about 1,500 people from the surrounding area are getting indirect employment from this factory. If seen in the same way, about 1,500 people are getting direct employment and about 4,500 people are getting indirect employment from the fertiliser factories of Gorakhpur, Sindri and Barauni. Talcher factory is also providing direct employment opportunities to 500 people while 1,500 people are getting indirect employment from this factory.

With the restart of these five old fertiliser factories, Bharat's annual urea production capacity will increase by about 63 lakh tonnes. With this, farmers across the country will continue to have adequate quantity of urea available for their crops. Farmers of Telangana, Andhra Pradesh, Tamil Nadu and Karnataka will especially benefit from the Ramagundam factory. The fertiliser factory of Gorakhpur will work to meet the urea needs of the farmers of Purvanchal. While the Sindri factory will supply urea to Jharkhand and adjoining states, the Barauni fertiliser factory will supply urea to the farmers of Bihar and its adjoining states. The Talcher plant will play an important role in meeting the urea requirements of Odisha and its neighboring states.

The commencement of these urea factories will help in bridging the gap between demand and supply of urea. Besides, small and medium enterprises will also develop around these factories and will supply goods and services locally to these factories. This will create an environment of entrepreneurship at the local level and strengthen the local economy.

Additional urea production of about 63 lakh tonnes from these factories will play an important role in making Bharat self-reliant in urea. Bharat has to spend foreign currency on importing urea from other countries. These new urea production units will also help Bharat save forex reserves.

❑ ❑ ❑

"Now, for self-reliance in urea production, Bharat is rapidly moving towards liquid Nano Urea. Nano Urea is a means for more production at less cost. Just imagine, the work that requires one bag of urea can now be done by a small bottle of Nano Urea."

"

This is a miracle of science, a miracle of technology, and due to this, farmers are freed from all the troubles of carrying urea bags, their hard work, transportation expenses and also the space to keep them at home. Now you go to the market, take ten things, put a bottle of Nano Urea in your pocket and your work is done.

"

4

Nano Urea: A Gamechanger

The Green Revolution of the 1960s had played a huge role in eliminating food crises from Bharat. Chemical fertilisers played an important role in making this Green Revolution a success. At that time, chemical fertilisers were considered a boon because of the way they increased productivity. With time, these chemical fertilisers also brought along with them some problems. This created issues like soil degradation, water pollution and air pollution, and people started considering chemical fertilisers as a bane, which once was considered as a boon during the Green Revolution.

In this regard, Prime Minister Shri Narendra Modi ji made an appeal to farmers and agricultural scientists in his Independence Day address from the Red Fort on August 15, 2019. The Prime Minister said, "I want to appeal to my farmer brothers today. I want to ask something from you. For my farmers, for my countrymen, this earth is our mother. As soon as we say Bharat Mata Ki Jai, there is a surge of energy within us. As soon as Vande Mataram is chanted, one gets the inspiration to sacrifice for this Mother Earth. We have a long history and tradition but have we ever got worried about the health of this Mother Earth? The way we are using chemicals, chemical fertilisers and pesticides, we are destroying our Mother Earth. As a child of this mother, as a farmer, I have no right to destroy my Mother Earth. I don't have the right to make my Mother Earth sad, I don't have the right to make my Mother Earth sick. It is going to be 75 years of independence. Respected Bapu has shown us the path. Can we reduce the use of chemical fertilisers in our fields by 10 per cent, 20 per cent, 25 per cent? Can we run a Muktikar campaign? You will see what a great service it will be to the country. What a big contribution you will be making in saving our Mother Earth."

Taking inspiration from this call of the Prime Minister, the Union Ministry of Chemicals and Fertilisers asked the fertiliser companies and other experts to work in this direction. Indian Farmers Fertilisers Cooperative Limited (IFFCO) developed Nano Urea under the guidance of the Ministry. Bharat has become the first country in the world to start the commercial production of Nano Urea. The production of Nano Urea at Kalol plant started on August 1, 2021, and till mid-January, 2024, 7.64 crore bottles have been produced and 7.23 crore bottles have been sold to farmers. Nano Urea particle is 55,000 times smaller than a particle of conventional urea. Whereas its surface area is 10,000 times more than that of a particle of conventional urea. With this, the plants will get the nutrients as per their requirement and they will be able to get better nutrition. Nano Urea is liquid urea. The amount of nitrogen contained in a bag of 45 kg conventional urea can be

Interesting Facts

The production of Nano Urea at Kalol plant started on August 1, 2021, and till mid-January, 2024, 7.64 crore bottles have been produced and 7.23 crore bottles have been sold to farmers.

Do You Know?

Bharat has become the first country in the world to start commercial production of Nano Urea.

supplied through just a half litre bottle of Nano Urea.

Before delving in detail into the features of Nano Urea and its benefits, it would be relevant to briefly understand the importance of urea in agriculture. Actually, plants of any crop require a total of 16 nutrients for balanced growth. Of these, nitrogen, potassium and phosphorus are called primary nutrients. Urea works to fulfill the nitrogen requirements of plants. Urea contains 46 per cent nitrogen. Adequate availability of nitrogen ensures proper growth of plants and also maintains their green color.

Now let us understand the current situation of urea use and its disadvantages. There is imbalanced use of urea in Bharat. The biggest reason for this is that the nitrogen use efficiency is only about 35 per cent. This means approximately 65 per cent of the nitrogen that plants should get from urea is being wasted. Soil is getting spoiled due to excessive use of urea and water and air pollution are also increasing.

Nano Urea is proving to be a revolutionary product towards solving these problems. Scientists involved in developing and testing Nano Urea claim that it can reduce the use of conventional urea by 50 per cent. The fertiliser use efficiency of Nano Urea is much better than conventional urea. Due to the large surface area of Nano Urea, the nutrients reach the plants slowly and the plants get better nutrition. The 65 per cent nitrogen that is wasted in the use of conventional urea will not be wasted by the use of Nano

Urea. Plants will get better nutrition.

A 50 per cent reduction in the use of conventional urea will have many benefits. The biggest benefit will be that of protection of soil health. Water and air pollution will be reduced. This means that Nano Urea is an eco-friendly product. With its use, agricultural productivity will also increase and the quality of products will also improve. Apart from this, the expenditure on urea subsidy will also be rationalised. The total fertiliser subsidy in 2022-23 was about Rs 2.5 lakh crore. Out of this, subsidy of approximately Rs 1.69 lakh crore was given only on urea.

Due to the use of Nano Urea, farmers are also benefiting financially. Farmers have to spend Rs 266.50 for 45 kg of conventional urea. Whereas a 500 ml Nano Urea bottle which provides equivalent nutrition to the crop is available for less than Rs 225. That means farmers are saving Rs 41.50 on every 45 kg conventional urea used. In this way, farmers are getting direct savings of about 16 per cent.

The expenses incurred by companies and farmers on transportation of urea bags are also saved. Transporting urea to the farm fields is a big challenge for farmers in rural and remote areas. Then there are storage issues for which companies and farmers have to spend a lot of money. Nano Urea will be able to solve these problems. Farmers' income will increase, and there would be a reduction in transportation and storage costs, etc.

Interesting Facts

Nano Urea particle is 55,000 times smaller than a particle of conventional urea. Whereas its surface area is 10,000 times more than that of a particle of conventional urea.

Do You Know?

There is imbalanced use of urea in Bharat. The biggest reason for this is that the nitrogen use efficiency is only about 35 per cent. This means that approximately 65 per cent of the nitrogen that plants should get from urea is being wasted.

Thus, the country is moving towards fulfilling Prime Minister Shri Narendra Modi ji's dream of doubling the income of farmers. Online sales and services have been started across the country through e-market from July, 2021 in which Nano Urea is being sold to farmers with free delivery facility.

On the instructions of the Central Government to technically empower the farmers, fertiliser companies have provided 2 lakh battery operated sprayers and twenty thousand power sprayers to the channel partners for spraying Nano Urea. Also, more than 350 drone pilots have been trained and this number is increasing every day.

About 5 lakh Nano Urea bottles have been exported by IFFCO in the last 2 years to more than 25 countries. Out of this, 3.06 lakh bottles have been exported to Sri Lanka, sixty thousand bottles to Nepal, forty thousand bottles to Suriname and Kenya. Recently, IFFCO has signed an agreement for export of Nano Urea to USA, which is being seen as a big step. Besides, more than 35 countries have shown an interest in Nano Urea and samples of Nano Urea have been made available to more than 20 countries, which has strengthened the possibilities of its export to other countries. IFFCO has signed MoUs with cooperative institutions of Brazil and

Argentina for setting up production plants and marketing of Nano Urea. In this way, the world's first indigenously manufactured Nano Urea is creating a distinct identity at the international level.

Diversion of conventional urea is also a big problem. This problem has reduced considerably due to neem coating of urea but the problem still persists. Let us try to understand this through some facts. Approximately 350 lakh tonnes of urea is consumed annually in Bharat. Out of this, we import about 75 lakh tonnes. About 280 lakh tonnes of urea is indigenously produced in the country. It is suspected that every year lakhs of tonnes of urea is diverted to other areas. This means that crores of rupees of subsidy being given on urea is wasted. Nano Urea will help in reducing these losses.

Along with this, Nano Urea will also reduce the import of conventional urea in Bharat. Currently, about 76 per cent of the gas used in urea production in Bharat is imported. As the production of Nano Urea increases, not only will the expenditure on urea import be reduced but the expenditure on gas import will also reduce significantly. This will lead to significant progress towards fulfilling Prime Minister Shri Narendra Modi ji's dream of "self-reliant Bharat" (Aatmanirbhar Bharat) in the fertiliser sector. Besides this, the country will also be able to save foreign currency and Bharat's trade imbalance will also reduce.

It is estimated that foliar application of Nano Urea with better

Interesting Facts

The amount of nitrogen contained in a bag of 45 kg conventional urea can be supplied through just a half litre bottle of Nano Urea.

Do You Know?

Soil is getting spoiled due to excessive use of urea. Besides, water pollution and air pollution are also increasing. Apart from this, its major disadvantage is that the greater the quantity of urea used, the greater will be the burden of subsidy on the government.

application of recommended dose of conventional urea gives a yield advantage of 3-8 per cent compared to conventional fertiliser application. These figures of improvement in productivity have emerged from the testing of Nano Urea on 94 crops in more than 11,000 farms, more than 20 agricultural research institutes/ universities and agricultural science centres in different states of the country. Besides this, the use of Nano Urea will also improve the quality of agricultural products. After the use of Nano Urea on wheat crop, it was found that the wheat grown using Nano Urea has more nutrients and minerals as compared to the wheat grown using conventional urea. The amount of protein was found to be approximately 6 per cent higher.

In Tamil Nadu, the use of Nano Urea through drones is being experimented on paddy crop. The initial results of this experiment are encouraging. The number of tillers has increased. Increase in tillers directly means that productivity will increase. Nano Urea is not only increasing the productivity but its use will also keep the soil healthy. Apart from this, nutritious grains will also be produced, which will ensure better health of the people.

For an agricultural country like Bharat, it is necessary to increase the production of Nano Urea. Other government companies working in the field of fertiliser should also engage in this work. IFFCO signed MoU with public sector undertakings like National Fertiliser Limited and Rashtriya Chemicals and Fertilisers for technology transfer of Nano Urea. This technology transfer by IFFCO was done without any royalty in the interest of the country. Many companies are moving in this direction to start commercial production.

With these MoUs and technological innovations, the production of Nano Urea will increase in the coming days. This will help in increasing the popularity of Nano Urea among the farmers in a short span of time and farmers will be able to adopt this new product faster. By 2025, production from eight units of Nano Urea will start. It is noteworthy that production has started in 3 of these units—Kalol (Gujarat), Phulpur (Prayagraj) and Aonla (Bareilly), Uttar Pradesh, which will make it possible to produce 17 crore Nano Urea liquid bottles in the year 2023-24. Other plants of Nano Urea Liquid are likely to be started in Deoghar (Jharkhand), Bengaluru (Karnataka), Trombay (Maharashtra), Nangal (Punjab) and Namrup (Assam) by the year 2024. When production of Nano Urea starts from all these production units, Bharat's annual Nano Urea production capacity will be 44 crore bottles. Obviously, by producing on such a large scale, Bharat will not only be able to meet its urea needs but will also be able to export Nano Urea to

Interesting Facts

Urea works to fulfill the nitrogen needs of plants. Urea contains 46 per cent nitrogen. Adequate availability of nitrogen ensures proper growth of plants and also maintains their green color.

Do You Know?

Scientists involved in developing and testing Nano Urea claim that it can reduce the use of urea by 50 per cent. The fertiliser use efficiency of Nano Urea is much better than conventional urea.

other countries.

The biggest challenge for any game-changing technology is its adoption by the common people on a large scale. Nano Urea is facing the same challenge. Nano Urea is a revolutionary product but faster the farmers of the country adopt it, the sooner we will see its positive impact in the agriculture sector. Necessary awareness programmes are being run to ensure that the farmers of the country adopt Nano Urea. IFFCO and other companies have undertaken extensive promotional programmes to create awareness about Nano Urea through PMKSKs and have reached an audience of 58 crore across Bharat with promotional resources. Due to these awareness campaigns, farmers are adopting Nano Urea on a large scale in a very short span. It is expected to increase further in the coming days. The Prime Minister has also worked to increase the level of awareness among farmers about Nano Urea by mentioning it in many of his addresses and on some occasions in his "Mann Ki Baat" programme. This is fast building up the confidence of farmers about Nano Urea.

During his tenure, Prime Minister Shri Narendra Modi ji has laid great emphasis on scientific research and development through the policies of his government. The development of Nano

Urea by the agricultural scientists of Bharat and its commercial production is the result of this visionary thinking of the Prime Minister. Apart from agricultural scientists, this success will also inspire scientists and researchers working in other fields for new research and development. Nano Urea will prove to be a revolutionary step and game changer product in the agriculture sector in near future.

❑ ❑ ❑

“

This Amritkaal is the time for all of us to do something for Bharat Mata. When the war of independence was going on, the generation born before 1947 got the opportunity to die for the country. They never left an opportunity to die for the country, but for us there can be no greater opportunity than this to live for the country.

”

"

If the soil health is deteriorated, if the health of our Mother Earth is not well, her fertility will also decrease. If the water is polluted, then there will be more problems. The farmer experiences all this in his everyday life. In such a situation, to increase the agricultural production and get good yield, we will have to create new systems in farming, adopt more scientific methods and more technology with an open mind.

"

5

Revolutionary Nano DAP

In 2021, Bharat became the first country in the world to start commercial production of liquid Nano Urea. In the very first year, farmers of the country adopted Nano Urea in place of conventional urea on a large scale. This resulted in considerable financial savings to the farmers and they used Nano Urea more efficiently than conventional urea. Encouraged by the success of Nano Urea, the Narendra Modi government at the centre quickly worked on the plan to introduce Nano DAP. In 2023, Bharat became the first country in the world to commercially produce Nano DAP.

IFFCO took decisive steps towards the production of Nano DAP. IFFCO has started production of Nano DAP at its Kalol plant from March 8, 2023. After this, two more plants will be started in Kandla and Paradip. Additionally, Coromandel International Limited (CIL) has also set up a state-of-the-art Nano DAP production facility in Andhra Pradesh with a manufacturing capacity of 4 crore bottles (1 litre each) per year.

Nano Urea and Nano DAP will play a major role in realizing the vision of "self-reliant Bharat" (Aatmanirbhar Bharat) in fertilisers.

Like Nano Urea, Nano DAP is also going to be a very useful and successful product for the farmers. DAP has great importance among the various fertilisers used in agriculture. After spraying,

DAP dissolves and provides phosphate and ammonium to the plants, which enables proper growth of the plants. Balanced use of DAP is very important for the proper development of plants.

Bharat is the largest consumer of DAP in the whole world. A total of 105.31 lakh tonnes of DAP was consumed in Bharat in 2022-23. In 2021-22 this figure was 92.64 lakh tonnes. Bharat also imports the maximum amount of DAP. Bharat's share in the total import of DAP in the world is 38 per cent. Bharat had imported a total of 65.83 lakh tonnes of DAP in 2022-23 and 54.62 lakh tonnes of DAP in 2021-22.

To reduce the import of DAP, the Modi government is paying special attention to increase its domestic production. The government aims to realize the dream of "self-reliant Bharat" in the coming years on the DAP front. Therefore, emphasis has also been given to both public and private companies to set up joint ventures in countries having raw materials of DAP. Nano DAP can play an important role in making Bharat self-reliant in terms of DAP.

Interesting Facts

IFFCO has started production of Nano DAP at its Kalol plant from March 8, 2023. Two more plants will be started in Kandla and Paradip. Additionally, Coromandel International Limited has also set up a state-of-the-art Nano DAP production facility in Andhra Pradesh with a manufacturing capacity of 4 crore bottles (1 litre each) per year.

Do You Know?

Encouraged by the success of Nano Urea, the Narendra Modi government at the centre quickly worked on the plan to introduce Nano DAP. In 2023, Bharat became the first country in the world to commercially produce Nano DAP.

The fertiliser use efficiency of the bagged DAP currently used is only 20 per cent. This means that 80 per cent of the DAP applied in the farm fields is wasted. Because of this, farmers have to apply more DAP in their fields than the crop requires. On one hand, this leads to imbalanced use of DAP, while on the other hand, farmers have to spend more money to buy more DAP than required. Wastage of 80 per cent of the total DAP used pollutes the environment too. Due to this, the productivity of soil and health of people is affected.

Nano DAP is going to play an important role in solving these problems. The size of Nano DAP is 10 to 30 nanometers. Such small particles have been coated with organic polymers. Its advantage will be that plants will get nutrients gradually through Nano DAP. This will improve fertiliser use efficiency. The fertiliser use efficiency of IFFCO's Nano DAP is more than 72 per cent.

In comparison to the amount of conventional DAP applied in one acre of land, a very less amount of Nano DAP will be required for the same land. In paddy crop, 50 kg (1 bag) of conventional DAP has to be used in one acre of land. In January, 2024, its total price is Rs 2,477. Out of this, the farmer has to pay Rs 1,350 and the government gives a subsidy of Rs 1,127. Whereas for paddy crop on the same land, only 500 ml of liquid Nano DAP would be required. For this, farmers will have to spend only Rs 600-650. That means farmers will get a direct benefit of Rs 700-750.

Currently, DAP comes in 50 kg bag. Instead of one bag of DAP, just using a 500 ml bottle of Nano DAP will do the trick. Because of this, transportation of fertilisers will be easier. Its storage will be easy for companies and farmers. At the same time, the expenses of companies and farmers on transportation and storage of DAP bags will reduce significantly.

IFFCO has also found in the study related to the use of Nano DAP that two applications of Nano DAP leads to best results. Better effect of Nano DAP was seen when used for the first time with seed coating. The second time it can be sprayed with Nano Urea after 30 days of seed germination. In this way, the plants on which Nano DAP was used twice showed better growth as compared to traditional DAP. Besides this, these plants also performed better on other parameters of germination and growth.

Nano DAP will reduce Bharat's DAP imports which will have a positive impact on the entire economy. Also, in the case of DAP, the country will be able to take steps towards realizing the dream of

Interesting Facts

Bharat is the largest consumer of DAP in the whole world. A total of 105.31 lakh tonnes of DAP was consumed in Bharat in 2022-23. In 2021-22 this figure was 92.64 lakh tonnes. Bharat also imports the maximum quantity of DAP.

Do You Know?

The size of Nano DAP developed by IFFCO is 10 to 30 nanometers. Such small particles have been coated with organic polymers. Its advantage will be that plants will get nutrients gradually through Nano DAP. This will improve fertiliser use efficiency. The Nano DAP developed by IFFCO has a fertiliser use efficiency of more than 72 per cent.

"self-reliant Bharat". One advantage of the Nano DAP will be that the monopoly of the countries in the international DAP market will be broken and gradually the international market of DAP will become more competitive.

❑ ❑ ❑

"The government will not let our farmers be troubled by urea prices. When a farmer goes to buy urea, he believes that it is Modi's guarantee."

"

The government is working sincerely to reduce the expenses of the farmers and help them in their time of need.

"

6

New Initiative of Urea Gold

Prime Minister Shri Narendra Modi ji gives a lot of emphasis on innovation. Under the leadership of Modi ji, Bharat has adopted innovation in many fields during his tenure of more than nine-and-a-half years. The fertiliser sector is also not untouched by this. During the tenure of more than nine-and-a-half years of Shri Narendra Modi ji as the Prime Minister, the dominance of innovation in the fertiliser sector has increased. Bharat became the first country in the world to start the commercial production of Nano Urea. The country also set the same record in the case of Nano DAP. Today, Bharatiya farmers are using Nano Urea and Nano DAP on a large scale. With its use, fertiliser efficiency and crop productivity is increasing. Bharat's nano fertilisers have been so successful in the country that now their demand is coming from other countries too and fertiliser companies are exporting them to other countries due to which farmers of those countries are also getting benefitted.

Now, in the fertiliser sector, a new innovation is being adopted in the form of "Urea Gold". Prime Minister Shri Narendra Modi ji launched it in Sikar, Rajasthan in the presence of 2 crore farmers who were connected to this event through PMKSK network. The work of developing Urea Gold has also been done by a government company. Rashtriya Chemicals and Fertilisers Limited, working under the Ministry of Chemicals and Fertilisers has achieved the distinction of developing Urea Gold.

Sulphur is an important macronutrient for enhancing crop yield and crop quality which is more so for oil seeds and pulses. In view of large imports of edible oil and pulses by Bharat, there is an urgent need to enhance Bharat's domestic production of these varieties which is possible with application of sulphur to the soil to enhance the crop quality and quantity. Alarmingly, Bharatiya soils are found to be hugely sulphur deficient and an urgent need is felt to effectively address this deficiency by means of adding supplementary sulphur inputs to the soil. The deficiency of sulphur in Bharatiya soils is bound to increase in the future as the requirement of sulphur far exceeds its application through all fertiliser sources. With realization of this factor began the quest for development of an effective and economical grade of fertiliser which could overcome the huge sulphur deficiency in the Bharatiya soils.

With the above objective in mind, a journey began in pursuit of indigenous development of an ideal grade of sulphur-carrying fertiliser blended with urea which is the most commonly used source of nitrogen. After a thorough examination of the various grades of sulphur containing fertilisers available in the "Fertiliser Control Order", it was felt that a nitrogenous-based fertiliser will have the best acceptability in terms of availability, affordability and efficacy. With the Government's proactive role in the revival of closed down urea plants at Gorakhpur, Sindri, Barauni and Ramagundam, a substantial quantity of more than 50 lakh metric tonnes (LMTs) per annum has been recently added to the existing urea manufacturing capacity of our country. Moreover, revival of the strategically important coal gasification-based urea manufacturing plant at Talcher is also in progress and will add additional 12.50 LMTs per annum. Sulphur, being a by-product of refining and smelting processes, is adequately available in the Bharatiya markets. Considering availability of these raw materials, it was felt that urea prills coated with sulphur could offer an ideal solution to address the need of nitrogen as well as sulphur to the soil.

With 17 per cent sulphur and 37 per cent nitrogen, sulphur coated urea would not only provide the important macronutrient sulphur apart from nitrogen to the soil but would also provide many allied benefits in terms of higher "Nutrient Use Efficiency" due to slow release of nitrogen, improved yield, better yield quality and reduction in soil leeching and water contamination.

Once the product was shortlisted, next challenge came in terms of its production. Though coating of sulphur over urea prills appeared easier on paper, when it came to actual production, several issues were encountered. However, with painstaking efforts by our scientists, engineers and technicians working in the state-owned Rashtriya Chemicals & Fertilisers, teething issues related to fluidity of molten sulphur, prill size of urea, nozzle size of sprayers, identification of right temperature, adequate pressure for uniform spraying of molten sulphur over urea prill, etc. were sorted out. The final product was a fine blend of yellow sulphur coating over the white urea prill with an appearance like Urea Gold.

Once the finished product was developed, pricing of the product was an important decision. It was felt that with the benefits of Urea Gold in terms of better productivity with lesser quantity and improved yield quality, the pricing should be kept at comparative levels to other fertiliser grades containing sulphur. However, considering the dire need to overcome the sulphur deficiency of Bharatiya soils and to provide an affordable fertiliser

Do You Know?

After coating of sulphur to urea, Urea Gold contains 37 per cent nitrogen and 17 per cent sulphur. The conventional urea that has been used till now contains 46 per cent nitrogen.

having the dual benefits of nitrogen as well as sulphur to the farmers, it was decided to keep the price of a 40 kg bag of Urea Gold at a highly subsidized rate of Rs 266.50 for the farmers. This value added urea need not be confused with the price or availability of the conventional neem coated urea which continues to be available in 45 kg bags at the existing price without any change.

In the quest to address sulphur deficiency in Bharatiya soils began the arduous journey of Urea Gold, and now when the objective is finally achieved, Urea Gold is ready to fulfill a long standing requirement of Bharatiya agricultural sector.

After coating of sulphur to urea, Urea Gold provides a comprehensive nutrition package for crops. With this, the production of specific crops can be increased by meeting their special needs. Urea Gold also increases the efficiency of nitrogen use. Due to this, nitrogen continues to be supplied to the crop for a long time. The crop remains green for a longer period. Obviously, this will have a positive impact on the productivity of crops. Farmers will benefit financially from this. Besides, farmers will get economic benefits in another way. Due to the increased efficiency of nitrogen use, farmers will need to use less amount of urea. Also, once the crop is fertilised, there will be no need to fertilise it again and again. This will also save farmers' money.

Generally, it is seen in villages that when farmers see that the color of their crops has started turning yellow, they apply fertiliser again. Now, due to the use of Urea Gold, the greenness of the crop will remain for a longer time, and then it is obvious that the farmers will not adopt the method of applying fertiliser again and again. This will not only provide financial savings to the farmers but will also improve the health of the soil due to rational use of fertilisers.

A few years ago, the Modi government launched neem coated urea. This not only increased the efficiency of urea use as fertiliser but also helped in curbing the misuse of urea for non-agricultural purposes. The efficiency of Urea Gold is even better than neem coated urea. It is a result of continuous efforts by the government towards increasing the efficiency of fertilisers, Urea Gold is getting available to crores of farmers in the country.

Urea Gold is not only capable of replacing conventional urea but also increasing its use will reduce the total urea usage. Due to better fertiliser efficiency, 25 per cent less usage compared to conventional urea also yields similar results. A study was conducted on this issue which revealed that crops are getting the same amount of nutrition by using 20 kg of conventional urea as they are getting by using only 15 kg of Urea Gold. Obviously, if less fertiliser has to be used in the form of Urea Gold as compared to conventional urea, then farmers will also get considerable financial savings. Overall, Urea Gold is going to be a boon for the farmers in the long run.

❑ ❑ ❑

"The government has also initiated the NAMO Drone Didi scheme. I have been informed that during the Viksit Bharat Sankalp Yatra, demonstration of more than 1 lakh drones were conducted. For the first time in the country's history, the public is being connected in this manner through technology on a mission mode. Currently, training for the use of drones is being provided primarily in the agricultural sector. However, in the coming days, its scope is expected to expand to other sectors as well."

"We stopped black marketing of urea by adopting 100 per cent neem coating of urea."

7

Neem Coating of Urea

Earlier, news of diversion of subsidised urea was a common thing in the country. When subsidised urea used to get recovered during surprise checks/raids, it would be concluded that urea got diverted illegally for non-agriculture sectors like plywood, resin industry, animal fodder industry, dairy and explosive sectors. In such situations, the challenge before the government was to stop this misuse. For this, the Central Government took steps at the administrative level through the states, but this alone was not enough to stop the diversion of urea.

In this background, the Prime Minister of Bharat, Shri Narendra Modi ji took the decision that the urea being used by farmers in Bharat needs to be neem coated. Neem coating of urea should be understood in simple terms. When urea which was being used conventionally is coated with neem oil, this is called neem coated urea.

There are many benefits of neem coated urea. Firstly, it prevents the diversion of urea. Secondly, the neem coating increases nitrogen use efficiency which has been a major problem in our fertiliser sector. In such a situation, neem coating of urea proved to be a successful experiment.

There is a deep thought process behind reaching this decision that urea should be coated with neem only. Our scientists did lot of research on the issue and studied the results of coating urea with neem and many other products. These studies revealed that when it comes to crop growth/yield or quality of grains, efficiency of nitrogen use, in every case, neem coating of urea emerged as the best option.

In the experiment regarding neem coating of urea, it was also revealed that while conventional urea was releasing nitrogen rapidly, neem coated urea slowed down the speed of nitrogen release. Due to the rapid release of nitrogen, plants were not able to utilize it properly. Also, the nitrogen which the plants were not able to use was polluting the underground water and the surrounding air. But these problems were solved due to the gradual release of nitrogen by the neem coated urea.

Actually, efforts regarding neem coating of urea had started in 2008. After receiving approval from the Cabinet Committee on Economic Affairs, the Central Government issued a notification on June 1, 2008, in which it was said that coating and fortification of fertilisers will be promoted in the country. Initially, the Central Government made a policy that fertiliser companies could do coating or fortification of fertilisers up to a maximum of 20 per cent of their total subsidized fertiliser production capacity. On January 11, 2011, the Central Government increased this maximum limit from 20 per cent to 35 per cent.

Interesting Facts

There are many benefits of neem coating of urea. Firstly, it has prevented the diversion of urea. Secondly, neem coating increases the efficiency of nitrogen provided to plants or crops through urea.

Do You Know?

Neem coating of urea should be understood in simple terms. When urea which was being used conventionally is coated with neem oil, this is called neem coated urea.

However, all these efforts did not give the desired results because only 35 per cent of the fertiliser was within this range. Because of this, farmers were not getting the benefits which they deserved. The government led by the visionary Prime Minister Shri Narendra Modi ji decided on January 7, 2015 that it would now be mandatory for all the fertiliser companies to do neem coating of 75 per cent of their total production capacity of subsidised urea. This paved the way for neem coating of the entire quantity of urea used by farmers in Bharat.

Encouraged by the positive impact this decision created among farmers and the fertiliser industry, the Modi government took a historic decision on May 13, 2015, that all the urea produced in the country will be neem coated. This means that urea could not be sold in Bharat without neem coating. Opposition parties baselessly criticized the government by saying that the government has not given time to the industry to prepare itself for this. Whereas the truth is that the government not only prepared the entire industry by making 75 per cent neem coating mandatory but also ensured that Bharat's urea industry was fully prepared for 100 per cent neem coating of urea.

After this decision was made by the Cabinet Committee on Economic Affairs chaired by Prime Minister Shri Narendra Modi ji, the Department of Fertilisers of the Union Ministry of Chemicals and Fertilisers issued a notification on May 25, 2015 for 100 per cent neem coating of urea. It was made clear that all the urea produced in Bharat should have neem coating compulsorily from September 1, 2015. Bharat also imports urea in large quantities. The Department of Fertilisers fixed December 1, 2015 as the date for complete neem coating of imported urea also. In this way, the Modi government ensured that 100 per cent urea available in the market from December 2015 was neem coated.

The amount of preparation the Central Government undertook before taking this decision can be gauged from the fact that the government had given the task to the then Department of Agriculture, Cooperation and Farmers Welfare to study the impact of neem Coated Urea on crops. In the final report prepared by the Agriculture Development and Rural Transformation Centre of Bengaluru, many benefits of neem coated urea were enumerated. The first advantage of neem coated urea was that it improves the quality of the soil. Besides this, it also saves the money that farmers

Interesting Facts

Our scientists and researchers studied the results of applying urea along with neem and many other products. This study revealed that whether it is a matter of crop growth/ yield or quality of grains, efficiency of nitrogen use, in every case, neem coating on urea grains emerged as the best option.

Do You Know?

In the experiment regarding neem coating of urea, it was also revealed that while conventional urea was releasing nitrogen rapidly, neem coated urea slowed down the speed of nitrogen release. Due to rapid release of nitrogen, plants were not able to utilize it properly.

have to spend on pesticides and weed management as neem coated urea controls insects and weeds in the crop.

This study also established that use of neem coated urea increased the productivity of all crops and its by-products. Scientific studies have established that farmers' income increases due to increase in crop productivity. This study also revealed that those farmers who grow sugarcane, soybean, paddy, jute and maize would benefit the most as the highest increase was recorded in the productivity of these crops.

The government not only conduct studies on the effects of neem coated urea but also engaged some independent agencies to conduct studies. A consultancy firm found in its study that the level of awareness among farmers regarding neem coated urea is slightly low. In this study it was found that only 54 per cent of the farmers knew that the urea they were using had been neem coated. But among the farmers who knew that urea had been neem coated, 89 per cent also knew what the benefits of neem coated urea were. Of these, 88 per cent farmers counted at least one benefit. 52 per cent of such farmers admitted that the use of neem coated urea has reduced the incidence of pests and weeds.

Twenty-nine per cent of the farmers included in the study believed that crop productivity had increased due to the use of neem coated urea. Also, there were 25 per cent farmers who accepted that neem coating of urea had improved its use efficiency. There were 15 per cent farmers who believed that the use of neem coated urea had improved the health of the soil. Also, 10 per cent farmers believed that neem coated urea had reduced their expenditure on pesticides. As the use of neem coated urea increased, farmers responded positively to it. Farmers also acknowledged that they are getting higher production even by using less amount of neem coated urea as compared to conventional urea due to slower release of nitrogen from neem coated urea.

The Fertiliser Department also took several steps to ensure that the work of 100 per cent neem coating of urea is ensured. All the companies producing urea were asked to install CCTV cameras in their production units. Besides, instructions were also given to install CCTV cameras at the ports where urea was being imported so that the neem coating of urea can be monitored. All the companies producing urea were instructed to purchase neem oil used in neem coating only from genuine neem oil producing companies. Along with this, instructions were also issued to urea producing companies and urea importing companies to add as a

Interesting Facts

The first advantage of neem coated urea was that it improved the quality of the soil. It also saves money that farmers have to spend on pesticides and weed management as it controls insects and weeds in the crop.

Do You Know?

The Modi government took a historic decision on May 13, 2015, that all the urea produced in the country will be neem coated. This means that urea could not be sold in Bharat without neem coating.

condition in their tender to mention the quantity of neem oil they would require as per their production or import capacity.

To ensure the quality of neem oil, details related to neem oil were recorded in the Fertiliser Control Order. To ensure that the urea is being coated with the correct quality neem oil, the central laboratories collected samples of urea from different urea production units and tested it.

Due to the decision taken by the Modi government to do 100 per cent neem coating of the urea used by the farmers of the country, the farmers are getting good crops and due to reduction in many of their expenses, their income is ultimately increasing. Also, neem coating of urea has significantly curbed the diversion of urea meant to be used in agriculture. Due to this, on one hand, the benefits of government subsidy on urea are reaching to the farmers, while on the other hand, the diversion of urea for non-agricultural purpose has been curbed. Overall, neem coating of urea has proved to be a boon for the farmers.

❑ ❑ ❑

"

Bharat's development is possible only when her villages are developed. The development of villages will create a developed Bharat. This is why the government is working to provide all those facilities in villages, which till now were available only in cities.

"

"Mark my words, my dear family members, whatever we will do in this period, the steps we will take, the sacrifices we will make, the decisions we will take one after the other, will be the golden glory of the country's history in the coming thousand years. The events occurring during this period are going to have an impact for the next thousand years."

8

Solution-oriented Chintan Shivir

On the theme of "Fertilising the Future", the Department of Fertilisers, Union Ministry of Chemicals and Fertilisers, organized a Chintan Shivir (Brainstorming Camp) in New Delhi on March 12 and 13, 2022. Before going to the main topics of this Chintan Shivir, it is necessary to understand the basic concepts of Chintan Shivir.

The meaning and importance of Chintan has been explained in our scriptures. In this regard, it would be absolutely correct to say that there has been a rich process of thinking in Bharat. When we look for the meaning of Chintan in the Hindi dictionary, we come to know that it means the feeling of thinking deeply, pondering, repeatedly meditating on something to understand or think about it or doing it in the mind. It is deliberation, meditation or repeated remembrance. All these meanings involve deep thinking. It would be better to call the process of thinking deeply about a topic from different perspectives as Chintan (contemplation).

Tattva Chintan has been of great importance in Bharat. Chintan paves the way for us to find the simplest solution to the most complex problems. Our religion and our scriptures have also underlined the importance of Chintan from time to time. For instance, after the war of Kalinga, Emperor Ashoka had introspected and contemplated.

Emperor Ashoka won the battle of Kalinga and after this victory he did deep Chintan and it changed his heart. The result of this contemplation was that once the great war winner became an important follower of Buddhism and started propagating peace and non-violence.

The influence of Chintan has been present in our scriptures and our behavior since ancient times. Kautilya's "Arthashastra" is an ancient and authentic work on governance. In this, Kautilya has given seven elements of the state. These seven elements are: Swami (the king), Aamtya (ministers), Janapada (the territory), Durga (a fortified capital), Kosha (the treasury), Danda (justice or force), and Mitra (ally). In this, Kautilya further writes that the king has three sources of power, among which his Council of Ministers is an important source.

Contemplation with enlightened people and the Council of Ministers holds great value. Kautilya considered collective thinking more important than the ability of an individual, i.e., a ruler. The tradition of contemplation which was envisioned by Kautilya has been followed in every successful political system.

As a minister, when I analysed this tradition of Chintan, I focused on how to bring the process of Chintan into public administration.

Interesting Facts

When we look for the meaning of Chintan in the Hindi dictionary, we come to know that it means the feeling of thinking deeply, pondering, repeatedly meditating on something to understand or think about it or doing it in the mind. It is deliberation, meditation or repeated remembrance. All these meanings involve deep thinking.

Do You Know?

On the theme of "Fertilising the Future", the Department of Fertilisers, Union Ministry of Chemicals and Fertilisers, organized a Chintan Shivir in New Delhi on March 12 and 13, 2022.

Chintan is extremely powerful as a tool. We thought deeply about using this powerful weapon of Chintan in public administration and this paved the way for the beginning of a new era.

When respected Prime Minister Shri Narendra Modi ji was working as the Chief Minister of Gujarat, he used to hold "Chintan Baithak" with the officials. But its form was slightly different. We adapted that format by making changes as per our needs and started the Chintan Shivir afresh.

When I was working as the Shipping Minister, I had organized two Chintan Shivirs of that ministry and organized another Chintan Shivir while being Minister of State in the Ministry of Chemicals and Fertilisers. As the Cabinet Minister of the Ministry of Health and Family Welfare I organized four such Shivirs and as the Cabinet Minister of the Ministry of Chemicals and Fertilisers, I organized one Chintan Shivir. We organize our Chintan Shivirs at locations that are away from our normal work locations so that all the officers and participants taking part in it can devote their full attention to the Shivir and contribute towards best outcomes.

While preparing for the Chintan Shivirs, we also keep in mind that at the place where it is organized, there should be accommodation arrangements for all participants so that when we hold a meeting, there is not only discussion among ourselves but also discussions and deliberations continue at the mutual level among the participants beyond formal meetings. The participants discuss openly with each other and share their experiences. This also marks the beginning of a new relationship between the participants that is helpful in their official relationships.

These Chintan Shivirs also act as "ice breakers". I have often seen that officers working in the same ministry talk to each other over the phone but are not familiar with each other face to face and do not have a personal touch and mutual relationship. There is only formal communication among these people and there is no system of informal communication among them. But when these officers meet in the Chintan Shivirs, they get to know each other and talk to each other, and an element of friendship and personal relationship is also added. Ultimately, the administration, system and society at large benefit from this.

In this context, I would like to mention the first Chintan Shivir of the Ministry of Shipping. In that Chintan Shivir, for the first time, our port officials were meeting Director General (DG) of Shipping face to face. This strengthened the mutual relationship between these officials. They learned and understood in detail about each other's tasks and responsibilities. They also understood how we can help each other and work together towards a common agenda. After this, these officers came and thanked me for making such arrangements in the form of a Chintan Shivir. These people said that

Interesting Facts

Chintan paves the way for us to find the simplest solution to the most complex problems.

Do You Know?

Chintan Shivir provides a platform where all participants can freely participate irrespective of the seniority in the government and administrative positions. In the Chintan Shivir, every officer gets an equal opportunity to speak.

the relationship between port and shipping is like a lock and key but they had never met each other in such a friendly atmosphere.

Chintan Shivir provides a platform where all participants can freely participate irrespective of the seniority in the government and administrative positions. In the Chintan Shivir, every officer gets an equal opportunity to speak. Suggestions are taken from all the participants. When the entire forum agrees on something, only then we start working in one direction with full force. Chintan Shivir provides an opportunity to all stakeholders to talk fearlessly and openly.

During Chintan Shivir, different sessions are organized on different topics or on different dimensions of the same topic. In these sessions, first a presentation is given on the topic and then all the participants come together for in-depth discussion. Each session is planned in such a way that there is ample time for discussion after the presentation.

It is generally seen in the Chintan Shivir that when such intense discussions take place during a session, not only the sessions last longer than their scheduled time but also the interest of participants in the detailed discussions remain intact. Discussion of topics is not limited to the scheduled session only, but discussion continues also during the intervals between the sessions or after the end of the day and before the beginning of the next day's session.

Chintan Shivir also works to strengthen mutual trust and relationships among all the participants. All the participants eat together during the camp. Apart from formal sessions, they talk openly among themselves on other issues also. We do yoga and exercise together in the morning. We participate in cultural programmes. Mutual participation in all these activities establishes a relationship of ease and mutual trust among the participants and this also brings ease in administrative work.

I have also seen that the goals set in the Chintan Shivir are accomplished in a short time because even before it is decided, all the stakeholders related to the subject discuss these subjects in detail on the platform of Chintan Shivir and then, after setting the target with mutual consent, they start working unitedly towards achieving it.

I have seen many issues which have been pending for years, but when they were taken up for discussion in the Chintan Shivir, the solutions became easy. In this process, when all the stakeholders sit on a platform and talk face to face, a consensus is reached within a few minutes around the solutions. Mutual doubts are resolved by knowing each other's thoughts face to face.

Interesting Facts

When respected Prime Minister Shri Narendra Modi ji was working as the Chief Minister of Gujarat, he used to hold "Chintan Baithak" with the officials.

Do You Know?

During Chintan Shivir, different sessions are organized on different topics or on different dimensions of the same topic. In these sessions, first a presentation is given on the topic and then all the participants come together for in-depth discussion.

Overall, Chintan Shivir is a platform where all the stakeholders come with the knowledge and experience of their respective fields and, through mutual participation, openly discuss the agenda.

In the background of this concept, when we organized a brainstorming camp on the theme of "Fertilising the Future", it was decided to discuss broadly five major topics. These are:

1. Reforms in Urea Policy
2. Global Fertiliser Scenario: Strategy and Way Forward
3. Integrated Plant Nutrition Management Act
4. Model Fertiliser Retail Outlet
5. One Nation, One Fertiliser

On the first day of the Chintan Shivir, the Managing Director of Chambal Fertilisers and Chemicals Limited gave a presentation on the topic of reforming the urea policy. He gave basic information about the entire urea industry and highlighted that for the last 15 years, making urea in Bharat is becoming cheaper than imported urea and this has saved the country about Rs 1.5 lakh crore. He also said that the government is clearing the subsidy payment backlog of companies, which is benefitting the sector.

He also gave many constructive suggestions for improvement in the urea sector like increasing our domestic production as much as possible. There were suggestions regarding energy saving and making a special policy for the new units to be installed in a certain period. Issues were raised for removing freight subsidy to simplify the entire process. These were discussions regarding pre-determined fixed subsidy for urea, Maximum Retail Price (MRP) to be decided according to the market conditions.

In this session, there were suggestions to start DBT 2.0 for giving fertiliser subsidy directly to the farmers wherein a "fertiliser card" can be issued to the farmers. By taking this card, farmers can encash their subsidy amount through POS machines installed at fertiliser centres.

In this session, various aspects related to gas supply were also discussed in detail. Most of the fertiliser companies supported the fact that the entire amount should be paid directly to the gas supplier. The reasoning behind this was that by doing so, the working capital requirement of fertiliser manufacturing companies would be reduced and the government would also be able to properly evaluate their impact on gas payments, fertiliser prices and subsidies.

Some important decisions were taken through the brainstorming process among all the stakeholders regarding reforms in the urea

Interesting Facts

We organize our Chintan Shivirs at a distant location from our workplace so that all the officers and participants taking part in it can devote their full attention to the Chintan Shivir and contribute to it with full participation.

Do You Know?

Chintan Shivir also works to strengthen mutual trust and relationships among all the participants. All the participants eat together during the camp. Apart from formal sessions, they talk openly among themselves on other occasions also. We do yoga and exercise together in the morning.

sector. To meet the spot gas requirements, the task of purchasing high pressure, high temperature gas will be done through Gas Authority of India Limited and Indian Oil Limited. It was also decided that the government would provide one-time help to those urea production units whose production is less than 6 lakh tonnes. It was also proposed to make direct payment to gas suppliers. For this, it was decided to make Gas Authority of India Limited an aggregator. Along with this, a decision was also taken that urea units will be classified on the basis of their production capacity/production cost. It was also decided to conduct further study on the subject of direct cash transfer through Aadhaar linkage of land.

A proposal to establish a Central Fertiliser University also came up in this brainstorming camp. It was decided that all aspects related to this would be studied. It was also decided that ICMR and the Fertiliser Department will jointly study the impact of deficiency in micronutrients on human health. A decision was also taken that fertiliser companies will create a fund for scientific research and development.

During the Chintan Shivir, the Managing Director of Indian Potash Limited gave a presentation in the session organized on the topic "Global Fertiliser Scenario: Strategy and Way Forward". He highlighted that the prices of natural gas have been increasing continuously for some time and due to this, the fertiliser industry is facing many challenges. He also gave details of the problems of fertiliser availability at the global level. It also came to light that in view of the fertiliser shortage, countries like China, Russia and Egypt are imposing restrictions on the export of fertilisers from their countries. In such a situation, the challenges for us are more complex because we are highly dependent on imports in terms of nitrogen, potash and P2O5 used as raw materials.

The participants in the Chintan Shivir gave many important suggestions regarding what should be the way forward for Bharat in these global circumstances. A consensus was reached among all the stakeholders to encourage the use of Nano Urea to reduce dependence on imported urea. In the case of phosphates, two suggestions came. The first suggestion was that more phosphoric

Interesting Facts

While preparing for the Chintan Shivir, we also keep in mind that at the place where it is organized, there should be accommodation arrangements for all the participants so that when we hold a meeting, there is discussion among ourselves but even when the meeting is not going on, discussions and deliberations continue at the mutual level among the participants.

Do You Know?

The goals set in the Chintan Shivir are accomplished in a short time. Because even before it is decided, all the stakeholders related to the subject discuss these subjects in detail on the platform of Chintan Shivir and then after setting the target with mutual agreement, they start working unitedly towards achieving it.

acid production units should be set up in Bharat itself. The second suggestion was these units should be set up as joint ventures in those countries where there is sufficient availability of rock phosphate. To solve the problem of disposal of gypsum, it was agreed to work on an option to explore the possibilities of increasing its use in the cement industry.

There were also suggestions in the Chintan Shivir that Bharat should enter into long-term agreements with major fertiliser producing countries to ensure its fertiliser security. Following this, the Modi government made long-term agreements with many countries. Fertiliser companies entered into long-term agreements with fertiliser companies from countries like Morocco, Senegal, Jordan, Oman, South Africa and Tunisia to ensure uninterrupted supply of fertilisers.

Consensus was also reached on increasing the use of SSP. It also came to light that at present this industry is working only at 55 to 60 per cent of its total capacity. It was agreed to increase it to at least 85 per cent in the coming days. It was also agreed that the production capacity of Nano DAP and Nano Urea needs to be increased significantly in the coming years. It was also agreed to promote research to explore the possibilities of alternative fertilisers. IFFCO was asked to go ahead and take lead in this work.

In this Chintan Shivir, there was a detailed discussion among various stakeholders on the topic of "Model Fertiliser Retail Shop" and "One Nation, One Fertiliser". It was agreed that all fertilisers should be marketed across the country under a single brand name "Bharat". It was launched by Prime Minister Narendra Modi in July 2023 in Sikar, Rajasthan in the presence of approximately two crore farmers. All these farmers joined this programme through the network of Prime Minister Krishi Samriddhi Kendra. It was also agreed in the Chintan Shivir that the model fertiliser retail shops will not only provide fertilisers to the farmers but also extend help in solving their other problems. Both these topics have been discussed in detail in the subsequent chapters.

❑ ❑ ❑

Interesting Facts

When these officers meet in the Chintan Shivir, they get to know each other and talk to each other and an element of friendship and personal connection emerges. Ultimately the administration, system and society benefit from this.

"

I have seen the structure of Pradhan Mantri Kisan Samriddhi Kendra and the model that has been set up. I really congratulate Mansukh Bhai and his team that it is not just a centre for buying and selling fertilisers for the farmer. Kisan Samriddhi Kendra is a complete centre for the farmer that connects him closely, answers all his questions and helps him in all his needs.

"

"

These centres will not only ensure fertiliser availability, but will also provide seeds, equipment, soil testing, every kind of information that the farmer needs at one place. Our farmer brothers and sisters, by visiting these centres, will get rid of all struggles to get basic inputs of agriculture.

"

9

One-stop Shop for Farmers: Pradhan Mantri Kisan Samriddhi Kendra

On October 17, 2022, the Prime Minister of Bharat, Shri Narendra Modi ji inaugurated 600 Pradhan Mantri Kisan Samriddhi Kendras (PMKSKs) from New Delhi. After this, the Honourable Prime Minister dedicated more than 125,000 PMKSKs to the nation on July 27, 2023. The Modi government aims to develop more than 2.5 lakh fertiliser sales centres across the country as PMKSKs. This work is to be completed in the next few months.

From May 26, 2014, Shri Narendra Modi took over as the Prime Minister of the country and since then, he has continuously shown his commitment through various policies and actions. New initiatives are being taken continuously under the leadership of the Prime Minister to improve the lives of farmers. During Modi ji's tenure of more than 9.5 years, many such steps have been taken which have strengthened and empowered the farmers of the country.

The Prime Minister has always worked with a total approach in the agriculture sector. Be it adopting new technology in agriculture, promoting the use of smart technology or providing better markets to farmers' produce, a lot of work has been done on all these fronts under his leadership. Therefore, on one hand, our agricultural production has reached at record levels, and on the other hand, farmers are getting better and fair prices for their produce.

During the tenure of the Prime Minister, the research in agriculture sector got a boost. For this reason, we became the first country in the world to start the commercial production of Nano Urea. Today, farmers of the country are using Nano Urea on a large scale. Also, at the appeal of the Prime Minister, the inclination of farmers towards natural farming is also increasing across the country. The level of awareness among farmers regarding the balanced use of chemical fertilisers is increasing.

In fact, the idea behind setting up the Pradhan Mantri Kisan Samriddhi Kendras is that these centres should act as a "one stop shop" for all the needs of the farmers. All the fertiliser shops working at village, block, sub-division and district levels are to be converted into Pradhan Mantri Kisan Samriddhi Kendras. New centres will also be developed as PMKSKs. To decide on the structure of these centres, the Government held extensive discussions with all the stakeholders. A detailed plan was worked out based on these deliberations.

Pradhan Mantri Kisan Samriddhi Kendras are empowering farmers in many ways. On the one hand, these model shops are working as fertiliser retail shops, and on the other hand, these centres are providing information and many useful facilities to the

Interesting Facts

On October 17, 2022, the Prime Minister of Bharat Shri Narendra Modi ji inaugurated 600 Pradhan Mantri Kisan Samriddhi Kendras simultaneously from New Delhi.

Do You Know?

The Prime Minister has always worked with a total approach in the agriculture sector. Be it adopting new technology in agriculture, promoting the use of smart technology or providing better markets to farmers' produce, a lot of work has been done on all these fronts under his leadership.

farmers under one roof. Farmers are able to get quality fertilisers, seeds and pesticides used in farming under one roof through these centres. Farmers have started getting useful information related to farming at these centres equipped with internet facilities. Besides this, facilities like soil and seed testing are also available at these centres.

There is also a plan to make agricultural equipment available at these centres, wherein small equipment is available for sale and big equipment is available for renting purpose. These centres are fast emerging as hubs for spreading awareness about "best practices" in agriculture sector.

Information about how farmers can avail the benefits of various schemes of the government is also being given through these centres. In order to ensure that farmers continue to get better products and services through these centres, arrangements have also been made for regular training for the operators of these centres.

There are currently more than 2.5 lakh fertiliser sales centres across the country. All these centres are to be developed as Pradhan

Mantri Kisan Samriddhi Kendra. The Government has planned to complete this work in a phased manner.

Some standards have been set to ensure that PMKSKs across the country look similar. The Department of Fertilisers had issued the Standard Operating Procedure (SOP) for opening PMKSKs across the country on August 24, 2022. The sign boards of all the centres will be of same type. The selection of colors will also be the same. Apart from Hindi, information will also be written in local languages and English on the boards of PMKSKs. The centres have been given the freedom to include photographs of local farmers. Space has been fixed for writing the information like name, GST number, address, etc., of these retail outlets. These centres can put the logo and name of their respective company.

The Pradhan Mantri Kisan Samriddhi Kendras in villages will be of about 150 square feet. There will be racks for the products at all these rural centres. There will be proper arrangement for sitting of farmers and other people. It is mandatory to have a Point of Sale (POS) machine at every centre. There will also be a QR code/bar code scanner. Apart from this, there will be arrangements for display of products. There will be a display board installed at every centre through which it will be known what the status of the stock is, what the MRP of the products is and how much subsidy is being given on any product. Apart from this, useful agricultural literature will also be available at every Pradhan Mantri Kisan Samriddhi Kendra. A

Interesting Facts

Honourable Prime Minister dedicated more than 125,000 Pradhan Mantri Kisan Samriddhi Kendras to the nation on July 27, 2023.

Do You Know?

The concept behind setting up the Pradhan Mantri Kisan Samriddhi Kendras is that these centres should act as a "one stop shop" for all the needs of the farmers. Under this, all the fertiliser centres working at village, block, sub-division and district level are to be converted into Pradhan Mantri Kisan Samriddhi Kendras.

chart related to the soil fertility of the area will also be displayed at each centre.

The size of the PMKSKs at the block and sub-division level will be about 200 square feet. These centres will have all the facilities mentioned earlier at the rural level centres along with some additional facilities as well. There will also be a smart television connected to the internet at all the centres at the block and sub-division level. This smart television will have many uses. Through this smart television, films related to best practices in the agriculture sector will be shown to the farmers visiting these centres. Films based on success stories of farmers and information about new technology and equipment used in the agriculture sector will also be screened through these smart televisions. These smart televisions can also be used to connect farmers with experts. Thus, apart from information on the usage of fertilisers, farmers will be able to solve their doubts related to agriculture with the help of experts. Special care has been taken to ensure that farmers get information in their local language so that they can use it properly.

These centres at the block and sub-division level will have the facility of a help-desk. This help desk will provide useful information to help the farmers. These centres will also be connected to the Common Service Centre so that farmers can get information on many more essential services through these centres. Soil testing facility will also be available at Pradhan Mantri Kisan Samriddhi Kendras at block and sub-division level. Besides this, facility to collect samples for testing of seeds and pesticides will be available at these centres. Through these centres, farmers can also rent agricultural equipment, drones, etc.

At the district level PMKSKs, apart from above facilities, some other important facilities will also be available. These centres will be of at least 2,000 square feet. Here all the facilities will be available on a larger scale than the village, block and sub-division level centres. There will be a relatively large size display installed here. There will be seating arrangements for more people. The facility of testing soil, seeds, water and pesticides will also be available at the district level Pradhan Mantri Kisan Samriddhi Kendras. There will be a Common Service Centre at the district level centres so that farmers can also avail other important services through these centres.

All types of quality fertilisers will be sold through PMKSKs, be it nitrogenous fertilisers or phosphatic fertilisers. Besides this, secondary and micro nutrients will also be available to the farmers through these centres. Alternative and organic fertilisers will

Interesting Facts

The Modi government plans to develop more than 2.5 lakh fertiliser sales centres across the country as Pradhan Mantri Kisan Samriddhi Kendras.

Do You Know?

To decide the structure of PMKSKs, the Government has held extensive discussions with all the stakeholders. On the basis of these detailed discussions, a plan has been made to establish and run these centres.

also be made available through these centres. The efforts of the Government are aligned to promote use of alternative fertilisers to the tune of at least 20 per cent through these centres.

The Government has also suggested to install ATMs at Pradhan Mantri Kisan Samriddhi Kendras wherever possible. A suggestion has been made to install solar panels at every centre so that these centres can meet their energy needs.

Understanding the local needs of the area where the Pradhan Mantri Kisan Samriddhi Kendra will be located, it will also act as a medium to provide appropriate guidance and advice to the farmers. These centres will be able to provide information about good agricultural practices to the farmers through agricultural experts of the states. Farmers will be told what type of soil they have and which crop will be more suitable. Information will also be given as to which fertilisers would be better to use according to the existing health of the soil.

Information about market prices of agricultural products will also be provided to farmers through Pradhan Mantri Kisan Samriddhi Kendras. The status of arrival of different agricultural product in the market will be shared through these centres. With these initiatives, farmers will be able to get better prices for their produce. Farmers will also get information about weather forecast through these centres. They will be able to take timely and appropriate decisions related to their farming and will be able to protect themselves against vagaries of nature.

To ensure that farmers continue to get quality products and proper services through these centres, the Government has also made arrangements for regular training for the operators of Pradhan Mantri Kisan Samriddhi Kendras. The work of training the operators of these centres will be done by the concerned fertiliser company every six months. Scientists, subject experts and retired agricultural experts of Krishi Vigyan Kendra will be called to provide training to the centre operators. Fertiliser companies will share the information and impacts of each training programme with the Department of Fertilisers of the Central Government. With this, the Central Government will keep a direct eye on the quality of these programmes and operational aspects of the centres.

The operators of Pradhan Mantri Kisan Samriddhi Kendra will also use modern technology to deliver correct information to the farmers. There is a plan to create a WhatsApp group by the name of "Kisan Samriddhi Group". In this, progressive farmers from the areas surrounding the respective centre will be added. Through this group, they will not only have meaningful dialogue among themselves but will also be able to access updated information

Interesting Facts

During the tenure of the Prime Minister, the research in agriculture has got a boost. For this reason, Bharat became the first country in the world to start commercial production of Nano Urea.

Do You Know?

These centres are also providing many information and facilities to the farmers under one roof. Farmers are able to get quality fertilisers, seeds and pesticides used in farming under one roof through these centres.

related to agricultural work. Weather forecast information will also be immediately sent to the farmers through Kisan Samriddhi Group. Market prices will also be shared among these groups.

Overall, PMKSKs will pave the way for prosperity of farmers. More than 1.6 lakh existing fertiliser sales centres have been converted into Model Pradhan Mantri Kisan Samriddhi Kendra by September 17, 2023.

Since the time when Prime Minister Shri Narendra Modi was the Chief Minister of Gujarat, he has been emphasising on direct communication with farmers in his governance policies. There were many areas of Gujarat where farmers had to face many problems. To solve these problems, Shri Narendra Modi ji, as the Chief Minister of Gujarat, communicated directly with the farmers of the concerned areas. He also inspired his ministers and officials to communicate directly with the farmers. This has remained an inspiration behind our efforts in fertiliser sector also.

If we talk only about the fertiliser sector, the Government has spent a subsidy of about Rs 2.5 lakh crore last year. It is given for the benefit of farmers. In such a situation, it becomes important for the government to have a channel of communication so that there can be direct communication with the farmers to understand their issues and find solutions. PMKSKs provide that platform of interaction.

Prime Minister Shri Narendra Modi ji also interacted with the farmers on the occasion of dedicating 125,000 Pradhan Mantri Kisan Samriddhi Kendras to the nation from Sikar, Rajasthan on July 27, 2023. Addressing the farmers on this occasion, Modi ji said, "Today 1.25 lakh PM Kisan Samriddhi Kendras have been started in the country. Crores of farmers will directly benefit from these PM Kisan Samriddhi Kendras built at village and block level. All these centres will pave the way for prosperity of farmers in true sense. In a way, these are one stop centres for farmers. You, farmer brothers and sisters, often have to go to different places for agricultural goods and other needs. Now you will not have to face such difficulties. Now at Pradhan Mantri Kisan Samriddhi Kendra, farmers will get seeds and fertilisers. Apart from this, farming related tools and other machines will also be available at this centre. These centres will provide every modern information related to farming to the farmers. I have seen that many times my farmer brothers and sisters suffer a lot due to lack of information about the scheme. These Pradhan Mantri Kisan Samriddhi

Interesting Facts

To solve the problems, as the Chief Minister of Gujarat, Shri Narendra Modi ji directly communicated with the farmers of the concerned areas. He also inspired his ministers and officials to communicate directly with the farmers. It was through these dialogues that a solution to the problems of the farmers emerged.

Do You Know?

A need was felt to develop some framework for regular communication with farmers on various issues related to agriculture and fertilisers. After considering different proposals, our government planned to develop a network of Pradhan Mantri Kisan Samriddhi Kendra across the country.

Kendras will now become a source of timely information about every scheme to the farmers."

The Prime Minister further said, "This is just the beginning and I urge my farmer brothers to also develop this habit, even if you do not have to buy anything related to farming, but if you have gone to the market, go to the Kisan Samriddhi Kendra in the area. Just look at what's going on. You must have seen that our mothers and sisters go to buy vegetables, but if they see a saree shop somewhere, they may not want to buy it but will visit the shop. Friends, by the end of this year, more than two-and-a-half lakh Prime Minister Kisan Samriddhi Kendras will be developed in the country."

These Pradhan Mantri Kisan Samriddhi Kendras are working as a "one stop centre" for providing essential information to the farmers along with fertilisers, seeds, pesticides and other agricultural inputs under one roof. These Kisan Samriddhi Kendras are being developed as centres of direct communication with farmers. In this context, there was a direct communication at my level with the farmers and operators of Pradhan Mantri Kisan Samriddhi Kendras on September 12, 2023. In this, I put forward the entire concept behind the PMKSKs and explained to the operators of these centres how we have to communicate directly with the farmers through these centres. I told the operators of these centres as well as the farmers that only through "Kisan Samvad" can rural prosperity be achieved. We have to understand that without "village development" our vision of "national development" will remain incomplete.

In my conversation, I asked this question to the operators of Pradhan Mantri Kisan Samriddhi Kendras, "Do you think these fertiliser shops can become a medium of change?" Everyone agreed that as a one-stop centre, Pradhan Mantri Kisan Samriddhi Kendra will not only be a fertiliser shop but will emerge as a centre of knowledge, dialogue and cooperation. We all have to work together in this direction.

In our country, the network of more than 2.5 lakh PMKSKs is being developed in such a way that these centres will work as the most important medium at the grassroots level to communicate with the farmers and spread awareness among them. In the coming days, Prime Minister Shri Narendra Modi himself will reach out to the farmers through these Prime Minister Kisan Samriddhi Kendras and the Prime Minister will be able to have direct, effective and productive communication with the farmers living in villages in every corner of the country.

❑ ❑ ❑

Interesting Facts

How to have direct communication with farmers? What is the solution for this? The Prime Minister interacted directly with farmers on many occasions and tried to understand the views of farmers and farming and show them the right path.

> One Nation, One Fertiliser, a plan to provide cheap and quality fertilisers to farmers under the Bharat brand, has started today.

"Now with One Nation, One Fertiliser, farmers are going to be freed from all kinds of confusions and better fertilisers are also going to be available. Now go to any corner of Bharat, urea will be sold with the same name, same brand, and same quality and this brand is Bharat! Now urea will be available in the country only in Bharat brand. When there will be only one brand of fertiliser in the entire country, the brand confusions will also end. This will also reduce the cost of fertiliser and it will be available in sufficient quantity quickly."

10

One Nation, One Fertiliser

The Prime Minister of Bharat, Shri Narendra Modi ji launched the Pradhan Mantri Bhartiya Jan-Urvarak Pariyojana (PMBJP)–"One Nation, One Fertiliser" scheme from the country's capital New Delhi on October 17, 2022. The task of supplying fertiliser in adequate quantity at the right time across the country is performed by the Fertiliser Department of the Central Government. This work is done with the cooperation of the states, and the entire system is managed on the basis of demand for fertilisers for every season. By maintaining regular communication with the fertiliser companies, necessary supplies are arranged as per the demand.

The Government spends huge subsidy on fertilisers so that sufficient quantity of fertilisers can be made available to the farmers of the country at affordable rates and they can easily do farming at low cost to ensure food security of the country.

Although the same fertiliser is manufactured by different companies, its nutrients are the same. For example, urea may be manufactured by different companies but the amount of nitrogen in it is the same (46 per cent). Similarly, no matter which company is making DAP, the quantity of its nutrients will remain the same. The quantity and proportion of different nutrients in fertilisers is decided under Fertiliser Control Order of 1985.

Farmers have little idea that irrespective of the brands of urea or DAP, their nutritional value remains the same. Because of this, it was observed that the inclination of farmers towards a particular brand of fertiliser increases in a particular area. This also happens because a particular company spends more money on its marketing. Also, a particular company has a strong network of retailers in a particular area.

There are many disadvantages of brand preference. Suppose the farmers of a particular area want urea of a particular brand, but it is not available in the local market while urea of another brand is available. Despite this, farmers would like to get urea of the brand of their choice. Because of this, they will not buy urea of other brands and an artificial situation of urea shortage is created at the local level. However, in reality there is no shortage of urea in that local market.

Due to the demand of a particular brand of fertiliser in a particular area, the fertiliser has to be sent from a distance to that area. This increases the subsidy burden of the government on transportation of fertilisers. This means the burden on the government treasury increases because criss-cross movement of fertilisers starts taking place. Since the Government gives subsidy on transportation of fertilisers to fertiliser companies, it does not matter to them how far they have to transport their fertiliser. It has

Interesting Facts

The Prime Minister of Bharat, Shri Narendra Modi ji launched the Pradhan Mantri Bhartiya Jan-Urvarak Pariyojana (PMBJP) – "One Nation, One Fertiliser" scheme from New Delhi on October 17, 2022.

Do You Know?

In the case of fertilisers, it is a fact that even if the same fertiliser is manufactured by different companies, its nutrients are the same. For example, urea may be manufactured by different companies but the amount of nitrogen in it is the same (46 per cent).

been observed that not bothering about the distance, companies give more emphasis on sending fertilisers to those areas where their market is better and their network of retailers is stronger. No financial burden of this entire process falls on the companies but it does fall directly on the government.

This can be understood through an example. A large part of the urea produced by Krishak Bharati Cooperative (KRIBHCO) or IFFCO in Uttar Pradesh goes to Rajasthan. This happens because the farmers of Rajasthan prefer urea of these brands. Whereas the urea produced by Chambal Fertilisers Limited in Rajasthan goes to Uttar Pradesh because in Uttar Pradesh, people like the urea of Chambal Fertiliser brand. The urea produced by IFFCO in Uttar Pradesh goes to Madhya Pradesh. Whereas the urea which is produced by National Fertilisers Limited in Madhya Pradesh is sent to Uttar Pradesh. The "One Nation, One Fertiliser" scheme aims to remove such discrepancies.

With the aim of overcoming these problems, "One Nation, One Fertiliser" scheme was started. Under this, there is now only one brand of fertilisers—"Bharat". Fertilisers of different companies are being sold all over the country with the name "Bharat Brand" and its logo.

If seen from the farmers' perspective, this step is ensuring better availability of fertilisers around them. Due to this, farmers are getting adequate quantity of fertiliser at the right time. Also, now there is no confusion among farmers about which brand of fertiliser would be better to use. The unnecessary or unhealthy competition between fertiliser companies is reducing. This will not only benefit the fertiliser companies but the entire fertiliser sector. The subsidy given by the government on freight transportation of fertilisers has decreased and is further decreasing. This will reduce the financial burden on the government treasury. Besides, the task of real-time monitoring of movement, availability and sale of fertilisers is becoming easier.

Several rounds of meetings were held with all stakeholders to prepare the "One Nation, One Fertiliser" scheme. Inputs from fertiliser companies were taken in great detail. In such a situation, it is wrong to say that after the introduction of this scheme, fertiliser companies will lose their identity. One-third of the front portion of any fertiliser bag is for the company to put their name and logo. In fact, if seen in its true sense, it is an example of Public Private Partnership, i.e., PPP model between the government and companies.

Here, I would like to highlight the initial efforts taken by the Department of Fertilisers in reducing the criss-cross movement. The movement of fertilisers was rationalised through scientific

Interesting Facts

In Bharat, the proportion of different elements in fertilisers is decided by the Fertiliser Control Order of 1985.

Do You Know?

Due to the demand of a particular brand of fertiliser in a particular area, the fertiliser has to be sent from a distance to that area. This increases the subsidy given by the government on transportation of fertilisers. This means the burden on the government treasury increases because criss-cross movement of fertilisers starts taking place.

and technical interventions in monthly supply plans. These efforts proved handy in the conceptualisation of the "One Nation, One Fertiliser" scheme which is benefitting crores of farmers in the country.

❑ ❑ ❑

"Imagine a scene in a village...wherein a sister of my self-help group is operating a drone and assisting in farming. Under this initiative, 15,000 drones will be provided to Namo Drone Didis. Their training has started and I am pleased to share that more than a thousand Namo Drone Didis have already been trained."

"The primary focus of our government is to modernize the rural economy and empower farmers. Therefore, the government is making efforts to strengthen small-scale farmers, reduce their agricultural expenses and ensure that they receive better prices in the market for their produce."

11

Fertiliser Minerals: Key to Aatmanirbharta

If Bharat has to take concrete steps towards achieving self-reliance (Aatmanirbharta) in fertiliser production, then first of all it is necessary that the minerals which are used as raw materials in making fertilisers, should be mined on a large scale in Bharat. The government led by Prime Minister Shri Narendra Modi ji has continuously worked in this direction. The results so far in this direction are encouraging and it is expected that progress will be even faster in the coming days.

Rock phosphate is required as raw material for the production of phosphatic fertilisers. We are largely dependent on imports for this. Rock phosphate and phosphoric acid are required to make DAP, SSP and NPK fertilisers. With the aim of reducing Bharat's import dependence on these minerals, indigenous mining has been accelerated.

The government has found in its studies that there are large reserves of low-grade rock phosphate in the areas around Udaipur in Rajasthan. From here, the process is on to increase the mining of rock phosphate to be used in the production of SSP. Several rounds of discussions have taken place to increase the mining of rock phosphate in the country. In one such meeting held on July 26, 2021, it was decided that the Department of Fertilisers will work with the Mining Department of the Central Government to increase the mining of phosphate and potash so that the companies producing fertilisers in Bharat can get raw materials from their own country. The outcomes at this stage are encouraging.

The Mining Department has released the list of mines that can be reserved for public sector undertakings under Section 17A of the Mines and Minerals (Development and Regulation) Act (MMDR Act), 1957. Through these steps, meaningful efforts are being made to increase the mining of rock phosphate and other raw materials used in making fertilisers in Bharat.

Besides this, efforts are on to ensure uninterrupted supply of raw materials from abroad. A Gujarat government company, Gujarat State Fertilisers & Chemicals (GSFC) has bought 19.98 per cent stake in Canada's Karnalyte Resources Inc. Under this, it is proposed to supply 3.5 lakh tonnes of high quality potash to GSFC every year. This will improve the availability of potash in Bharat and the fertiliser needs of farmers can be met in a timely manner.

We are completely dependent on imports to meet our potash requirements. But now significant progress is made towards mining of potash and glauconite minerals. For their mining, the concerned agencies of the Central Government have identified

Interesting Facts

The government has found in its studies that there are large reserves of low-grade rock phosphate in the areas around Udaipur in Rajasthan. Government is working towards increasing the mining of rock phosphate so that it can be used in the production of SSP.

Do You Know?

Rock phosphate is required as a raw material for the production of phosphatic fertilisers. Bharat is largely dependent on imports for this. Rock phosphate and phosphoric acid are required to make DAP, SSP and NPK fertilisers. With the aim of reducing Bharat's import dependence on these minerals, mining has been accelerated.

many mines in Rajasthan, Madhya Pradesh and Uttar Pradesh. To start their mining and expand it properly, the Mining Department has started the process of fixing royalty under the MMDR Act, 1957. The Department of Fertilisers has given the list of identified mines to the public sector undertakings. After this, FAGMIL (FCI Aravali Gypsum and Minerals [India] Limited) and GSFC have shown interest in mining of potash. There are plans to start its mining domestically in the coming days. Once potash mining starts in Bharat, import dependence in terms of MOP will reduce and the country will be able to move towards fulfilling Prime Minister Shri Narendra Modi's dream of "Aatmanirbhar Bharat".

To strengthen Bharat's position in terms of availability of fertiliser minerals, fertiliser companies have also bought stakes mines abroad. The Modi government has continuously encouraged fertiliser companies for this. Two fertiliser companies have together purchased 37 per cent stake in Jordan Phosphate Mines Company. These two companies are India Potash Limited and IFFCO's subsidiary Kisan International Trading FZE. A joint venture is being set up with Morocco for the supply of phosphoric acid through pipeline. Under this, a plan has been made to supply five lakh tonnes of phosphoric acid annually. Through all these efforts, Bharat's resolve to ensure "fertiliser security" and "fertiliser self-sufficiency" is being strengthened.

❑ ❑ ❑

"When the purchasing power of the poor increases, the business power of the middle class increases. When the purchasing power of the village increases, the economic system of the town and city runs at a faster pace. This is an interconnected cycle. We want to move forward by further strengthening it."

“

The bag of urea which costs Rs 266 in Bharat is priced around Rs 800 in Pakistan, around Rs 720 in Bangladesh, around Rs 2,100 in China and around Rs 3,000 in USA. Our government will ensure that the farmers are not affected by rising international prices of Urea. Whenever a farmer goes to buy Urea, he is assured of Modi's guarantee.

”

12

Capacity Expansion in SSP

Single Super Phosphate (SSP) is a phosphatic fertiliser that is used extensively by farmers. After DAP, the most commonly used phosphatic fertiliser is SSP. Farmers use SSP on a large scale because it contains three nutrients like phosphorus, sulphur and calcium. Apart from this, many micronutrients are also present in it. This provides proper nutrition to the crop and increases productivity.

Considering the importance of SSP for agriculture, it is important to ensure that adequate quantity of SSP is available as per the needs of the farmers. Besides this, it should also be ensured that the quality of SSP being sold in the market is proper. There is also a need to ensure that there is regular supply of SSP in sufficient quantity in the market. We are also continuously working on the plans to expand the capacity of SSP production.

Rock phosphate is required as raw material for the production of SSP. Along with promoting the mining of rock phosphate domestically, the Modi government has also worked continuously to import rock phosphate from other countries as per the need.

The urgent need for capacity expansion of SSP production is even more because the farmers of the country are using SSP as an alternative to DAP. SSP is cheaper than DAP. In 2021, the trends of SSP replacing DAP were seen. Data from October 1, 2021, to

December 31, 2021, shows that SSP sales in the country increased by 56 per cent, whereas a decline of 21 per cent was recorded in the sales of DAP.

The Government has planned that only the units producing minimum 50,000 metric tons of SSP will be included in the nutrition-based subsidy system. To ensure quality, it has been planned to take the services of NABL Labs and Mobile Labs. A digital module will also be developed to follow quality assurance processes. It is the commitment of the Modi government to ensure quality fertilisers for the farmers.

Experts who have deep understanding of agriculture and fertiliser sector also believes that SSP and urea together will work as a more effective fertiliser than DAP, provided that SSP's quality is assured. This will provide additional nutrients like calcium and sulphur to the crops. In the mixture of urea and SSP, urea will be present in the form of urea calcium sulphate and nitrogen release will be slower in comparison to conventional urea. This will be more beneficial for the crop as nitrogen use efficiency will increase.

Currently, one tonne of imported DAP costs farmers around Rs 27,000. With this much DAP, the soil gets 180 kg nitrogen and 460 kg phosphorus. If seen in this way, the soil gets a total of 640 kg of primary nutrients. This mixture will provide more

Interesting Facts

Single Super Phosphate (SSP) is a phosphatic fertiliser which is used extensively by farmers. After DAP, the most commonly used phosphatic fertiliser is SSP.

Do You Know?

The Modi government has taken several policy decisions to promote the production of SSP in Bharat. Work has been done to widen the scope of mining of rock phosphate used in SSP production. Besides, work has also been done to simplify its production.

nutrients to the soil. With this mixture, the soil will get 184 kg nitrogen, 460 kg phosphorus and 316 kg sulphur. Overall, this mixture will provide 960 kg of primary and secondary nutrients to the soil, one-and-a-half times more primary and secondary nutrients! This may come as a surprise to many people. But this will prove to be very beneficial for the farmers. In addition to providing one-and-a-half times more primary nutrients, this mixture will also provide 604 kg calcium to the soil. The mixture of urea and SSP will also fulfill the needs of the soil for micronutrients like zinc and boron.

Farmers will make financial savings with mixture of urea and SSP. Farmers will be able to use this money to fulfill their other needs. Some will be able to provide better education to their children while others will use the money saved to provide better health services to their elderly family members. This savings will also reduce agricultural costs and improve farmers' income. The country will move forward on the path of self-reliance in fertiliser sector. Prime Minister Shri Narendra Modi ji's dream of "self-reliant Bharat" will be fulfilled by increasing the production of SSP in the coming days.

Overall, SSP is surely going to contribute positively to the fertiliser sector of the country. At the same time, it is going to provide many benefits to the entire country and the farmers on a large scale in the entire ecosystem.

❑ ❑ ❑

“We are making conscious efforts in a scientific manner to ensure that the best quality seeds are available to the farmers. In the last 7-8 years, more than 1,700 varieties of such seeds have been made available to the farmers, which can fulfill their purpose and remain adaptable even in these changing climatic conditions.”

"Due to NAMO Drone Didi initiative, the income of self-help groups is set to rise, nurturing their self-reliance. This initiative not only enhances the financial well-being but also cultivates a renewed sense of self-confidence among the sisters in the village, ultimately proving to be exceptionally beneficial for our hardworking farmers."

13

Fertiliser Crisis Management

In the last two years, there have been many problems regarding the availability of fertilisers at the international level. There were many reasons for this, over which we had no control. Despite this, under the visionary leadership of Prime Minister Shri Narendra Modi ji, we managed to avert this fertiliser crisis very well and did not let the farmers of the country face any shortage of fertilisers. During this period, it was also ensured that the farmers of the country do not have to pay more for fertilisers and there is no financial burden on them.

In fact, the demand for fertilisers across the world increased by 6 per cent in 2021. In the previous ten years, it had never happened that the demand for fertilisers increased by 6 per cent in any one year. In 2021, nitrogen demand increased by 5 per cent, phosphorus demand increased by 6.8 per cent and potassium demand increased by 9.1 per cent. This created an imbalance in the entire global market of fertilisers.

According to experts, this demand increased because many countries increased their investment in agricultural production to ensure food security. But the result of sudden increase in demand was that fertiliser prices started rising globally. Compared to January, 2021, the price of urea increased from $286 per tonne to around $1,000 per tonne in December, 2021. During this period, the price of DAP increased from $404 per tonne to around $900 per tonne. Whereas the price of MOP increased from $230 per tonne to $445 per tonne.

While the demand for fertilisers was increasing in the global market, the supply was affected drastically. Due to Covid-19, agriculture in many big countries was affected in 2020. After this, many countries like China and Russia imposed some restrictions on the export of fertilisers from their countries so that there is additional availability of fertilisers domestically and farming is done properly. As a result, the supply of fertilisers in the international market got affected.

In such a situation, there was a clear challenge before us of how to ensure timely supply of fertilisers in adequate quantity without increasing the prices. We could handle 2021–22 Rabi season well through better management. We were hoping that the prices of fertilisers would come down. But the situation further worsened due to US and EU sanctions on Belarus. After this, when the war between Russia and Ukraine started, the entire global market of fertilisers got disturbed. The supply of nitrogen, phosphorus and

Interesting Facts

Compared to January, 2021, the price of urea increased from $286 per tonne to around $1,000 per tonne in December, 2021. During this period, the price of DAP increased from $404 per tonne to around $900 per tonne. Whereas the price of MOP increased from $230 per tonne to $445 per tonne.

Do You Know?

Under the leadership of Prime Minister Shri Narendra Modi, Bharat managed the fertiliser crisis well and did not let the farmers of the country face any shortage of fertilisers. During this period, the Government of Bharat also ensured that the farmers of the country do not have to pay more for fertilisers and there is no financial burden on them.

potassium was badly affected. Potash's arrival in Southeast Asian countries reduced and ammonia supply in Morocco was affected, as a result of which the supply of phosphates from Morocco to Bharat was hampered.

Russia and Belarus together supply 40 per cent of our MOP requirements. Besides this, both supply 20 per cent of our DAP and NPK. After Covid-19, there was a surge in demand for agricultural products and fertilisers globally. In 2020–21, China exported 30 lakh tonnes of DAP to Bharat. In the beginning of 2021, China started curtailing the export of DAP and then in July–August, 2021, China curbed the export of DAP. US and EU sanctions on Egypt worsened the situation. Due to all these reasons, the increase in the prices of fertilisers globally picked up further.

We managed this crisis brilliantly. The entire situation was continuously monitored at my level under the guidance of the Honourable Prime Minister. The appropriate steps were taken from time to time. From the Kharif 2021 season onwards, regular communication and coordination was established with the fertiliser industry at the ministerial level.

A four-point strategy was adopted. Firstly, we entered into long-term supply agreements with fertiliser producers outside Russia and Belarus. Secondly, we tried to change the pattern of fertiliser consumption. By reducing the use of DAP, efforts were made to increase the use of NPK and SSP and increase their production at the domestic level. In states like Punjab, Haryana and Rajasthan, where DAP was predominantly used, the use of NPK fertilisers increased.

Thirdly, we established a close coordination with the states, and micromanagement of the supply of fertilisers was done. The Fertiliser Department set up a 24-hour control room. Its job was to establish coordination between State Governments and fertiliser companies so that there is no shortage of fertilisers in any part of the country.

The fourth step in this direction was to make farmers aware about fertiliser use efficiency so that farmers use fertilisers rationally. On the basis of this four-point strategy, the Government managed the global fertiliser crisis and saved its farmers and agriculture from the chaotic situation elsewhere in the world.

Interesting Facts

By reducing the use of DAP, the use of NPK and SSP was encouraged and their production was increased at the domestic level. In states like Punjab, Haryana and Rajasthan, where DAP was predominantly used, the use of NPK fertilisers increased.

Do You Know?

The demand for fertilisers across the world increased by 6 per cent in 2021. In the previous ten years, it had never happened that the demand for fertilisers increased by 6 per cent in any one year. In 2021, nitrogen demand increased by 5 per cent, phosphorus demand increased by 6.8 per cent and potassium demand increased by 9.1 per cent.

During this time, I encouraged all the major fertiliser companies to enter long-term fertiliser supply agreements with other countries and major companies there. I visited many such countries and played my part in concluding long-term fertiliser supply agreements. The Government took the help of its embassies located in different countries to make this happen. The result of all this was that in a situation where many countries were not getting fertilisers in sufficient quantities, fertiliser supplies to Bharat continued in adequate quantities at reasonable rates. As a result, our agriculture did well during these challenging times, and we could ensure our food security.

Just to narrate my experience of visiting various countries to get fertiliser supplies. I realized that there is a massive change in thinking of these countries about Bharat. These countries have now understood that this is the "New Bharat" led by Prime Minister Shri Narendra Modi ji, to whom they cannot "dictate" their terms but they will have to maintain a respectful "dialogue". This was the reason that during these visits the Modi government was successful in making such agreements through which fertiliser supplies were ensured. We were successful in curbing the arbitrariness of some countries in determining the prices of fertilisers at international level. During the tenure of the Modi government, such agreements are signed for fertiliser supply, which are fetching us supplies at rates suitable to us rather than binding to us.

The global fertiliser crisis has prepared us to face any such situations in the future. We have aligned all axes to ensure the country's "fertiliser security" and "food security."

We have started working at every level of research, production, supply, sale and consumption. The Modi government is continuously working to bring more efficiency in the entire process by bringing reforms at all these levels in the fertiliser sector. Information about the steps being taken at all these levels is given in detail in different chapters of this book.

❑ ❑ ❑

"After Covid, a new global order, a new geopolitical equation is emerging. All definitions of this equation are changing. My brothers and sisters, you will take pride in the fact that today the strength of my 1.4 billion fellow countrymen is visible in shaping the changing world. You are standing at a decisive juncture."

"Today Bharat is becoming the voice of the Global South. Bharat's share in global economy, global supply is increasing. I am confident that the role played by Bharat has guaranteed the world's stability."

14

Vasudhaiva Kutumbakam: Lending a Helping Hand

Bharat has been a country that believes in "Vasudhaiva Kutumbakam". Therefore, whenever other countries have faced any need, the Government of Bharat led by Prime Minister Shri Narendra Modi has stood by for help. When Covid-19 hit the whole world, Bharat not only fulfilled its own requirements of essential medicines but also supplied essential medicines to other countries of the world. Similarly, Bharat also played a major role in ensuring fertiliser security of its neighboring countries. By supplying adequate quantities of fertilisers to Nepal, Bhutan and Sri Lanka, Bharat worked to help the agriculture and farmers of these countries as well.

If we take the example of Nepal, the relationship between Bharat and Nepal is very old. This relationship is cultural and social first. Later a commercial dimension was added to it. The mutual relations between the people of both the countries are very deep. This is the reason why the governments of both the countries work with mutual coordination on many fronts. Both the countries also work together on many international forums like SAARC, BIMSTEC, etc.

Nepal is a landlocked country without any access to sea. Nepal's major imports and exports go through Bharatiya ports. Agriculture is the mainstay of Nepal's economy. Whenever the Government of Nepal has requested us to supply fertilisers for its agriculture, we have immediately addressed those. On February 28, 2022, a long term MOU was signed between the governments of the two countries under which Bharat will supply urea and DAP to Nepal. Bharat had supplied a total of 1.5 lakh tonnes of fertiliser to Nepal in 2021–22. Under the new agreement, the quantity of fertiliser supplied from Bharat to Nepal will increase from the current level to 2.1 lakh tonnes by 2025–26. Apart from this agreement, Bharat has also supplied 87,500 tonnes of urea, 20,000 tonnes of DAP and 7,500 tonnes of MOP to Nepal in the year 2022–23.

Similarly, we have deep-rooted relations with Bhutan. In 2021, we exported 2,000 tonnes of urea to Bhutan. Similar arrangements on the lines of the ones with Nepal are being worked out so as to take care of fertiliser needs of Bhutan in the long run.

Similarly, the Sri Lankan government requested for immediate supply of 65,000 tonnes of urea. In view of the internal situation in Sri Lanka and the immediate fertiliser requirement, we immediately approved the supply of 65,000 tonnes of urea. Rashtriya Chemicals and Fertilisers Limited was entrusted with the responsibility of supplying 65,000 tonnes of urea to Sri Lanka and they completed the work within the stipulated time.

Interesting Facts

Bharat had supplied a total of 1.5 lakh tonnes of fertiliser to Nepal in 2021–22. Under the new agreement, the quantity of fertiliser supplied from Bharat to Nepal will increase from the current level to 2.1 lakh tonnes by 2025–26.

Do You Know?

The Sri Lankan government requested for immediate availability of 65,000 tonnes of urea through Bharat's Ministry of External Affairs. In view of the internal situation and the immediate fertiliser requirement in Sri Lanka, the Bharatiya government immediately approved the supply of 65,000 tonnes of urea to its neighboring country Sri Lanka.

Bharat helped three neighboring countries at a time when it was paramount for us to ensure fertiliser security internally in the wake of the global fertiliser crisis. This helped the three neighboring countries in saving their agriculture and livelihood of the farmers.

IFFCO is supplying Nano Urea to more than two dozen countries including USA. It is noteworthy that the demand for Nano Urea made in Bharat is coming from many countries of the world and this demand is continuously increasing.

❑ ❑ ❑

Gorakhpur fertiliser factory in its new avatar.

Production also resumed at Barauni fertiliser factory in Bihar.

Production was started at Sindri fertiliser factory in Jharkhand.

Bharat is growing rapidly on the path of self-reliance in urea production.

Production of neem coated urea started at HURL fertiliser factory in Sindri, Jharkhand.

Prime Minister Shri Narendra Modi inaugurated the fertiliser factory in Ramagundam, Telangana.

Ramagundam fertiliser factory is now fully operational.

The efficiency of fertiliser factories is increasing with the use of modern technology.

Import dependence is decreasing through capacity expansion in urea production.

State-of-the-art technologies have been adopted in fertiliser production.

The "One Nation, One Fertiliser" project was launched by Prime Minister Shri Narendra Modi.

Through the "One Nation, One Fertiliser" scheme, fertilisers are now being made available under a single brand.

Prime Minister Shri Narendra Modi inaugurated the Kalol unit of IFFCO Nano Urea.

Farmers are getting many benefits due to increased use of drones in spraying revolutionary Nano Urea.

IFFCO transferred technology of revolutionary Nano Urea to RCF and NFL.

On July 27, 2023, the Prime Minister dedicated more than 1,25,000 Pradhan Mantri Kisan Samriddhi Kendras to the nation from Sikar, Rajasthan.

Farmers are getting many facilities under one roof at Pradhan Mantri Kisan Samriddhi Kendras.

Union Minister of Chemicals and Fertilisers Dr. Mansukh Mandaviya and Union Minister of State for Chemicals and Fertilisers Shri Bhagwant Khuba interacting with the operators of Prime Minister Kisan Samriddhi Kendra and the beneficiaries of this scheme on September 12, 2023.

Continuous review meetings are being held to make Pradhan Mantri Kisan Samriddhi Kendra a one-stop shop for farmers.

The Central Government has also made a plan for regular and necessary communication with the farmers through Pradhan Mantri Kisan Samriddhi Kendras.

Modi government is committed to creating a network of Pradhan Mantri Kisan Samriddhi Kendras across the country.

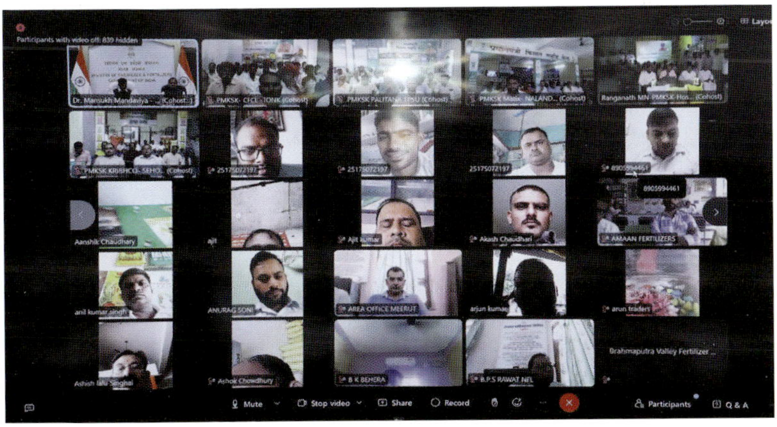

Operators of Pradhan Mantri Krishi Samriddhi Kendras and beneficiaries of this scheme shared their experiences.

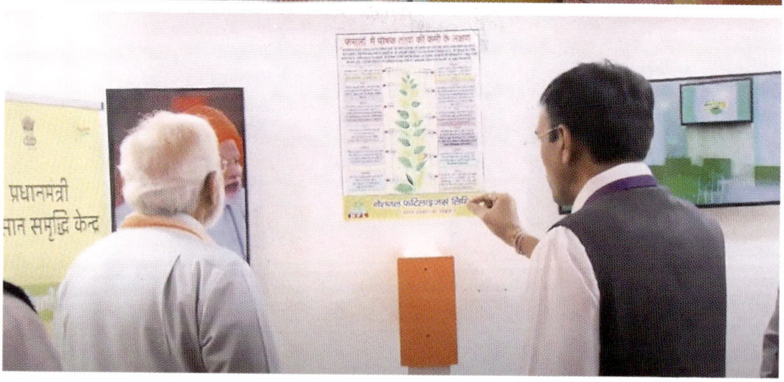

Prime Minister Shri Narendra Modi ji himself visited Prime Minister Kisan Samriddhi Kendra and saw the products and services available at these centres.

The Department of Fertilisers of the Central Government, in collaboration with states and other agencies, formed flying squads and took strict action against black marketing, smuggling and diversion of fertilisers.

Long-term agreements signed with the world's leading fertiliser producers.

Bharat's food security ensured through global agreements for supply of fertilisers.

Through international agreements, the Modi government ensured that Bharat's farmers get regular supplies of fertilisers at reasonable prices.

Bharat ensured fertiliser security in the country by entering into long-term agreements.

When crisis-hit Sri Lanka asked for fertiliser help from Bharat, the Modi government did not delay in sending fertiliser.

Bharat has made long-term agreements with many countries to ensure uninterrupted supply of fertilisers in the country.

World's Lowest Fertiliser Price in Bharat

Country	Urea	DAP
USA	₹ 2,364	₹ 3,282
China	₹ 1,512	₹ 2,126
Pakistan	₹ 773	₹ 3,158
Bangladesh	₹ 1,012	₹ 789
Nepal	₹ 1,120	₹ 3,440
Bharat	₹ 266	₹ 1,350

Note: This price is for a 50 kg bag. The prices of all countries except Nepal are of August, 2023, while the rate of Nepal is of March, 2023.

"Natural farming also provides an important way to solve future challenges. For this too, we are experiencing a lot of awareness across the country today. Farmers are working on natural farming on a large scale in Gujarat, Himachal Pradesh and Andhra Pradesh as well as in UP and Uttarakhand. In Gujarat, plans are being made in this regard even at the district and gram panchayat level."

"In the past years, the way natural farming has got new markets and the way it has been encouraged, the production has also increased manifold."

15

Marching Towards Alternative Fertilisers

Prime Minister Shri Narendra Modi ji is continuously presenting the concept of "Waste to Wealth" in front of the country. In this context, the Prime Minister talks about the four R's. These four R's are: Reduce, Reuse, Recycle and Repurpose. In order to realize this vision of the Prime Minister, the Department of Fertilisers took an important step and decided in October 2021 that now subsidy will be given on potash made from the waste of sugar mills. The Department of Fertilisers worked for this at various levels for a long time and due to these efforts this decision was taken.

There is a need to understand this whole thing in a very simple language. Actually, after crushing sugarcane, two products are formed. On the one hand, sugar is made from sugarcane juice. On the other hand, ethanol is also produced from sugarcane juice. After the process of ethanol formation is completed, a thick red-colored liquid is formed. The disposal of this liquid is required. To deal with this problem, with the efforts made in the last five years, on the initiative of Prime Minister Shri Narendra Modi ji, we brought this into the realm of subsidy.

This new initiative of the Ministry of Chemicals and Fertilisers is going to raise new hopes for disposal of inventories of red liquid. The Modi government has decided that subsidy will be given on potash being made through this liquid. This new potash made from molasses will be completely organic. The potash made from molasses will be an alternative fertiliser and the way the Prime Minister wants to increase the use of alternative fertilisers across the country, the potash made from molasses will prove useful in achieving that goal.

In the Cabinet meeting chaired by Prime Minister Shri Narendra Modi ji in October, 2021, it was decided to bring potash made from molasses under the Nutrient Based Subsidy (NBS) system. By including potash made from molasses as 25th grade in the NBS system, a historic step was taken. Already registered under the Fertiliser Control Order, this potash is now made available to the farmers of the country at a reasonable price. The production and use of potash made from molasses is accelerating in the country. In the last two years after this decision, the production of this type of fertiliser has also increased and its use is also increasing.

Sugarcane mills will also benefit from potash made from molasses. There are reports of poor financial health of sugarcane mills from many states of the country. Due to their poor financial condition, sugarcane mills are not able to make timely payments to the farmers for their sugarcane. To address such a situation, these mills will get additional income from potash made from molasses. The molasses which has been a waste for the mills and whose disposal has become a headache for them, is becoming a source of

Interesting Facts

This new potash made from molasses will be completely organic.

Do You Know?

Taking an important step, the Department of Fertilisers decided in October 2021 that now subsidy will be given on potash made from waste of sugar mills. The Department of Fertilisers worked for this at various levels for a long time and due to these efforts this decision was taken.

additional income for the sugarcane mills.

When the income of sugarcane mills increases, farmers will also benefit from it. These mills will be able to buy sugarcane from farmers at a better price than before. Also, when the income of these mills increases, they will also be able to make timely payments to the farmers. The second benefit to the farmers from this will be that they will be able to get cheap potash. This will reduce their costs in agriculture. This will surely help in achieving Prime Minister Shri Narendra Modi ji's goal of doubling the income of farmers.

This potash is completely organic as it is made from natural sources. This will strengthen the efforts of the Modi government to increase the use of organic fertilisers in the country. Experts in the fertiliser sector believe that potash made from molasses will also ensure better soil health. When the health of the soil improves, its fertility will remain longer and its productivity will also increase. The quality of crops grown on good soil will also improve, which will also help in ensuring better human health. It means potash made from molasses has brought many benefits.

Another advantage of potash made from molasses will be that

it will reduce the import of potash. In the last five years, Bharat has imported an average of 40 lakh tonnes of potash every year. Currently, Bharat is 100 per cent dependent on imports for its potash needs. By increasing the production of indigenous potash made from molasses, the country's expenditure on import of potash will reduce. Bharat will save its forex reserve. Ultimately, with this, the country is moving towards fulfilling Prime Minister Shri Narendra Modi's dream of "self-reliant Bharat" (Aatmanirbhar Bharat).

Cow dung has been in use as an agricultural fertiliser since ancient times. Before the beginning of chemical fertilisers, animal dung used to be collected by every family living in the village at one place and then it used to be collectively dumped in a field. Gradually, when the use of chemical fertilisers increased in the country, the trend of using cow dung as fertiliser decreased.

Actually, earlier, cow dung was not only used in the fields as fertiliser, but it was used as a source of cooking gas. But with time, as the use of chemical fertilisers increased in the fields, the use of LPG cylinders also increased in the kitchen. Due to this the use of cow dung decreased.

The Modi government started the GOBARdhan Yojana with the objective of promoting solid and liquid waste management by positively impacting village cleanliness and generate wealth and energy from cattle and organic waste. Under this, arrangements

Interesting Facts

Bharatiya government data for 2022 shows that the number of livestock in Bharat was 535 million. This shows how wide the possibilities are for using cow dung in agricultural purposes.

Do You Know?

There is a provision to make alternative fertiliser not only from cattle dung but also by using other agricultural residues and organic waste.

have been made for market development assistance.

Actually, GOBARdhan Yojana is based on the whole of government approach. Different ministries work on different parts of this scheme. Apart from the Ministry of Chemicals and Fertilisers, Ministry of Drinking Water and Sanitation, Ministry of Petroleum and Natural Gas, Ministry of Renewable Energy, Ministry of Animal Husbandry and Dairy, etc., are also playing their roles in this scheme at different levels. There is a provision to make alternative fertiliser not only from cattle dung but also by using other agricultural residues and organic waste. Apart from fertilisers, work is also being done to make biogas/CBG/bio-CNG from these residues. Separate schemes for all these products have been made under this scheme and remarkable progress is being made on every front.

The nodal department of this scheme is the Drinking Water and Sanitation Department. The Department of Drinking Water and Sanitation, being the nodal department, has developed a GOBARdhan-Unified Registration Portal. This portal has been developed for Biogas/CBG/Bio-CNG units. Operators of biogas/CBG/Bio-CNG units can register themselves by visiting this portal. Also, if anyone wants to start a new unit of Biogas/CBG/Bio-CNG, then they can also register themselves by visiting this portal.

Financial assistance can be provided under the Nurtient Based Subsidy regime only after registering on this portal. This financial assistance has been arranged by the Union Ministry of Chemicals and Fertilisers. To avail financial support from other ministries, it is mandatory that the concerned unit is registered on this portal.

Along with approving the plan of economic cooperation, CCEA also approved the proposal that a steering committee will be formed under the chairmanship of Secretary (Fertiliser). This committee comprises members of at least a Joint Secretary level officer from the Department of Expenditure, NITI Aayog, Department of Agriculture and Farmers Welfare, Department of Drinking Water and Sanitation, Ministry of New and Renewable Energy, Ministry of Housing and Urban Affairs, Ministry of Petroleum and Natural Gas, Department of Animal Husbandry and Dairying.

This committee will decide what should be the method of distribution of financial assistance and what guidelines should be prepared for it so that the right benefits reach to the right people. This committee was formed on July 18, 2023 and the committee has also started working. The first meeting of this committee was held on August 1, 2023 and a detailed discussion took place regarding the preparation of guidelines.

Overall, GOBARdhan Yojana and the market development assistance being provided under it are going to benefit farmers and mother nature at many levels. The cost of agriculture will reduce through GOBARdhan Yojana. Compared to expensive chemical fertilisers, organic fertilisers will be available to farmers from their own resources at a lower cost. Obviously, when agricultural costs

Interesting Facts

Earlier, cow dung was not only used in the fields as fertiliser. By making gas from cow dung, it was used as cooking gas and food was cooked with it.

Do You Know?

How cow dung can become a wealth for the farmers and a means to increase their income and at the same time ensure environment friendly farming—the Modi government started the GOBARdhan Yojana with these objectives. Under this, there are certain provisions for market development assistance.

will reduce, the income of farmers will also increase.

GOBARdhan Yojana will not only reduce the agricultural cost but will also bring many other benefits. If organic fertilisers are used in farming instead of chemical fertilisers, it will promote environment-friendly farming. Reducing the use of chemical fertilisers and increasing the use of organic fertilisers will improve the health of Mother Earth. If the health of the soil improves, the quality of the agricultural products will also be better. Water pollution will also reduce. This will have a positive impact on the entire environment and will strengthen environmental protection efforts. Overall, GOBARdhan Yojana is a scheme which is going to have a positive impact on a very large scale.

❑ ❑ ❑

"

In our country, governments have come and gone, numerous schemes have been devised, extensive discussions have taken place, grand promises have been made, and from all those experiences, from all the contemplation, I felt that the most crucial task that deserves attention in our country is ensuring that the government's plans, their purpose, and the work they are intended for, reach their destination without any delays or complications at the right time.

"

"The government has introduced a scheme called "NAMO Drone Didi", often referred to as "Namo Didi" by people in rural areas. This campaign aims to provide training to the sisters of around 15,000 self-help groups at the beginning of the initiative. These trained "Namo Drone Didis" will then be given drones. In the villages, they will perform tasks like spraying pesticides, applying fertilisers, inspecting crops, and checking if water has reached the fields—all tasks that can now be done by drones."

16

Result-oriented PM-PRANAM

Modi government is pro poor, pro farmers. Its clear imprint is visible in every decision of the Modi government from May 26, 2014 till now. The Prime Minister has continuously worked towards the progress of farmers. Whenever any challenge came for the farmers, the Prime Minister came forward and addressed it successfully. It is explained in a separate chapter of the same book that when there was a huge increase in the prices of fertilisers all over the world, at that time also Modi ji ensured that the burden of rising prices should not fall on the farmers. By increasing the subsidy, the Prime Minister not only ensured the availability of fertilisers to the farmers at lower prices, but he also ensured that there is no problem in the supply of fertilisers in the country.

Taking forward his commitment towards farmers and farming, Prime Minister Shri Narendra Modi ji has started the PM-PRANAM scheme for farmers. The full name of this scheme is PM Programme for Restoration, Awareness, Nourishment and Amelioration of Mother Earth. The main objective of this scheme is to increase the use of alternative fertilisers among farmers. Also, balanced use of chemical fertilisers should be encouraged. Presently there is situation of imbalanced use of fertilisers. In this, imbalanced use of urea is especially visible. Because of this, many types of problems are being faced. The health of the soil is deteriorating, pollution is increasing. To overcome these problems, it is necessary to promote balanced use of fertilisers across the country.

The scheme was announced in the budget for financial year 2023–24. This announcement clearly shows the commitment of the Government towards sustainable practices in the field of agriculture and conservation of natural resources.

The tendency to adopt alternative fertilisers will increase among farmers. Under this scheme, a provision has been made that states and Union Territories will be given incentives for the use of alternative fertilisers. Apart from this, a provision has also been made in this scheme to encourage balanced use of traditional fertilisers. Whichever state and union territory will reduce the use of chemical fertilisers and make good economic savings, they will be given financial grant as incentive from the Central Government. There is an atmosphere of enthusiasm clearly visible among states and Union Territories regarding the adoption of alternative fertilisers and many are actively participating towards this.

A special feature of the PM-PRANAM scheme is that it has a provision for adopting such technologies which can promote both the production and use of alternative fertilisers. Adoption

Interesting Facts

The scheme was announced in the budget for financial year 2023–24. This announcement clearly shows the commitment of the government towards sustainable practices in the field of agriculture and conservation of natural resources.

Do You Know?

Taking forward his commitment towards farmers and farming, Prime Minister Shri Narendra Modi has started the PM-PRANAM scheme for farmers. The full name of this scheme is PM Programme for Restoration, Awareness, Nourishment and Amelioration of Mother Earth.

of these technologies will bring efficiency in both production and utilization. Under PM-PRANAM, a provision has been made to spend 70 per cent of the total fund allocation in creating assets. This includes setting up production units at village, block and district levels, so that production of alternative fertilisers can start at the local level. This will bring many benefits. The availability of alternative fertilisers at the local level will improve. Once the supply system becomes stronger, the demand will also increase. This will ultimately improve the health of Mother Earth. Also, increasing the production of alternative fertilisers at the local level will create many new employment opportunities.

30 per cent of the total amount allocated under PM-PRANAM will be used to give incentives to farmers and other rural people. Under this, incentives will be given to those farmers who will reduce the use of traditional fertilisers by using alternative fertilisers. This will make those farmers feel honoured who will adopt better and nature-centric farming methods. Keeping this in mind, the target has been set in the scheme that the government has to work to encourage environment-friendly agriculture. Doing

this will reduce farmers' dependence on chemical fertilisers and will help in environmental protection.

For the environment to remain healthy, it is important that the Mother Earth remains healthy. For this, the most important thing is that there should be balanced use of fertilisers in agriculture. One of the basic objectives of the PM-PRANAM scheme is to promote balanced use of fertilisers. Under this scheme, emphasis is being laid on developing a long-term thinking on how to maintain the health of Mother Earth and ensure that productivity is not affected.

Various studies have been conducted which highlight the ill effects of imbalanced use of chemical fertilisers. The lesson of this study is also that the balanced use of fertilisers is very important to maintain soil health and productivity and avoid saturation of any kind.

The ideal ratio of N:P:K use is 4:2:1. But today the existing situation is 7.7:3.1:1. This has been the situation for the last several years. There is lack of awareness among farmers regarding balanced use of fertilisers. Because of this, they do not even realize that they are unknowingly harming the health of the soil and the environment as well as their own health. Today everyone, including farmers, have to understand that the Mother Earth as

Interesting Facts

The proposal to start the PM-PRANAM scheme came in the National Conference on Agriculture for Rabi Campaign organized in September 2022. Since then the Government has continuously worked on it. It was first announced in the budget and later it was launched.

Do You Know?

The main objective of this scheme is to increase the use of alternative fertilisers among farmers. Also, balanced use of chemical fertilisers should be encouraged. Presently there is situation of imbalanced use of fertilisers, especially of urea.

well as the environment and human health are all interconnected. If the situation goes awry on even one of these fronts, everyone will suffer. To change this situation, State Governments and Union Territories should also work together with the Central Government. With this vision the Modi government came up with the PM-PRANAM scheme.

Mother Earth has always provided abundant sources of livelihood to mankind. It is the need of the hour to promote more natural methods of farming and balanced/sustainable use of chemical fertilisers. Promotion of natural/organic farming, alternative fertilisers, nano fertilisers and bio fertilisers can help in restoring the fertility of our Mother Earth. All the states should jointly promote organic farming so that soil fertility is maintained and the ill effects of pesticides on human life can be reduced. If all the states and Union Territories work step by step in this direction along with the Central Government, then definitely the PM-PRANAM scheme will prove to be a scheme that will bring revolutionary changes in the agriculture sector of the country.

❑ ❑ ❑

"How much easier the farmer's life is going to be with kisan drones. The drone can provide data on soil quality, type of fertilisers required, how much irrigation is needed, what the nature of the disease is and what crop protection products are needed and it can give you the right guidance."

“

If fertiliser is to be sprayed, the drone sprays it only in the area where it is needed, and in right doses. This will also prevent wastage of spray and fertiliser and my farmer brothers and sisters will also be saved from the chemicals that they are exposed to.

”

17

Technological Interventions to Enhance Efficiency

In this era, it is difficult to imagine any area of life without technology. Today modern technology is being used in all aspects, whether big or small. This is making life easier and at the same time efficiency is also increasing. The use of technology is producing better results with less resources. Also, due to modern technology, it has become possible today that the benefits of any facility can reach more and more people and a wider area in a shorter time.

Bharat's fertiliser sector is also proactively adopting new technologies. This is increasing the efficiency of the entire fertiliser sector and farmers are also benefiting from it. The development of Nano Urea and Nano DAP and adoption of new technologies related to it can be taken as an example. Bharat became the first country in the world to commercially produce Nano Urea. Similarly, in the case of Nano DAP, Bharat set an example for the whole world.

Compared to conventional fertilisers, Nano Urea and Nano DAP are liquid fertilisers, and used through spraying. It takes a lot of time for farmers to spray with traditional sprayers and it also requires a lot of hard work. Given that, the Modi government decided to leverage drone technology for spraying liquid fertilisers.

There are many benefits of using this technology. First of all, it takes less time to spray with a drone. The second advantage is it reduces the cost of spraying. The third advantage is it creates new employment opportunities at the local level.

Drones are already used in the agricultural sector in developed countries of the world. This has brought revolutionary changes in the agricultural system of these countries. In recent years, on the initiative of the government, the use of drones in the agriculture sector has increased rapidly. Today, drones are being used in the agriculture sector not only for spraying fertilisers and pesticides but also for soil assessment, crop maintenance, production estimation, loss assessment in case of disaster and payment of insurance claims, etc.

There are approximately two-thirds of the people in our country who are dependent on agriculture in some form or the other. Therefore, if skill and efficiency increases in the agricultural sector, it will not only benefit the agrarian population but the entire economy.

Some of the benefits of using drones in fertiliser spraying were briefly mentioned earlier. Now let us understand this in a little more detail. If drones are used for spraying fertilisers, then fertilisers are sprayed over a large area in a very short time. If we take the example of traditional spraying through sprayer, then the drone can spray the same amount of fertiliser on the same area and crop in less hours compared to the traditional technique.

Interesting Facts

The use of technology is producing better results with fewer resources. Also, due to modern technology, it has become possible today that the benefits of any facility can reach to more and more people in a shorter time.

Do You Know?

Compared to conventional fertilisers, Nano Urea and Nano DAP are liquid fertilisers, and used through spraying. It takes a lot of time for farmers to spray with traditional sprayers and it also requires a lot of hard work. In such a situation, the Modi government decided to use drone technology for spraying liquid fertilisers.

Obviously this will save a lot of time for the farmers. Besides this, farmers will also save labour.

Spraying fertilisers using drones is also an economically beneficial deal for farmers. At present, they are able to spray fertilisers and pesticides in their fields by hiring drones for much less money than the wages they have to pay for spraying liquid fertilisers and pesticides through sprayers. Saving money for farmers simply means that agricultural costs are decreasing. The direct result of reduction in agricultural costs will be increase in farmers' income and farming will become more profitable than before.

This will also lead to creation of drone entrepreneurship at the local level. New employment opportunities are opening up for many youths. The youths are buying drones and renting them out for spraying purpose in farmers' fields.

IFFCO is buying 2,500 drones. Farmers will use these for spraying Nano Urea and Nano DAP. The government has also planned to provide drones to some selected rural entrepreneurs free of cost so that more and more farmers can spray fertilisers and pesticides on their crops using drones at the local level. Under this scheme of the government, an arrangement has been made that once the spraying work is completed in the targeted area, the ownership of the drones and three-wheeler electric loader will be given to the rural entrepreneurs so that they can use it in future also and its benefits can reach more and more farmers.

IFFCO will also do the work of providing necessary training for spraying with drones at the local level. It is making arrangements to train those youths who are interested in this work. The trained youths, to whom the drones will be provided by the government, will work to provide the service of spraying fertilisers through drones to the farmers at a concessional rate. Many fertiliser companies are pitching in similar activities. The drone has become the centre of attraction during Viksit Bharat Sankalp Yatra. The NAMO Drone Didi as highlighted in subsequent pages is a revolutionary step under the visionary leadership of Prime Minister Shri Narendra Modi ji.

❑ ❑ ❑

After the formation of the government in 2014, our government started working with a new approach. We simultaneously paid attention to both the day-to-day problems of the country's farmers and how to modernize agriculture and preparing it for future needs.

"

We used to hear a lot about modern farming in developed countries; the farmers there are so prosperous. We studied in depth what revolution has taken place in the agriculture sector across the world, what changes have taken place, what new initiatives have taken place and how it has been linked with the economy. After this we set our goals and started working on all of them together.

"

18

Farmer-friendly Innovations

Prime Minister Shri Narendra Modi ji places a lot of emphasis on innovation in every sector. Since he became Prime Minister on May 26, 2014, there is no sector in which he has not laid emphasis on innovation. Its positive results are also visible in all areas. During Modi ji's tenure of more than nine-and-a-half years, the fertiliser sector has also not lagged behind in terms of innovation. Over the years, intensive scientific research has led to many such innovations, through which not only commercial production of new fertilisers have started but also farmers are using them now on a large scale.

While interacting with farmers on December 25, 2020, Prime Minister Shri Narendra Modi ji underlined the importance of innovation in the agriculture sector and said, "After forming the Government in 2014, our government started working with a new approach. We simultaneously paid attention to both, addressing the day-to-day problems of the country's farmers and also how to modernize agriculture, and prepare it for the future needs. We used to hear a lot that farming in a particular country is so modern, the farmers there are so prosperous. We studied in depth what revolution has taken place in the agricultural sector across the world, what changes have taken place, what new initiatives have taken place and how it has been linked with the economy. After this we set our goals and started working on all of them together. We worked with the aim that the expenditure on farming by the farmers of the country should be reduced, i.e., the input cost should be reduced as well as the expenditure should be reduced. Schemes like soil health card, neem coating of urea and installation of solar pumps were taken up one after another to reduce their input cost. Our government tried to ensure that farmers have a better crop insurance cover. Today, crores of farmers in the country are getting the benefit of PM Fasal Bima Yojana.

In the long-term studies done by the country's top agricultural research institutes on the use of fertilisers, it has been found that by increasing the use of fertilisers, the production of food grains has increased by 25 to 50 per cent. In a study of the period from 1950 to 2021, it has been found that in terms of nutrients, the consumption of fertilisers worldwide has increased from 31 million tonnes to 188 million tonnes and in Bharat, the consumption has increased manifold, i.e., almost 50 times from 7 lakh tonnes to 33 million tonnes.

This is the reason why the annual growth rate in Bharat's fertiliser sector has remained more than double the world's annual

> **Interesting Facts**
>
> **During Modi ji's tenure of more than nine years, the fertiliser sector has also not lagged behind in terms of innovation. Over the years, intensive scientific research has led to many such innovations, through which not only commercial production of alternative fertiliser products has started but also farmers are using them now on a large scale.**

Do You Know?

It is a matter of pride for every Bharatiya that on the lines of Nano Urea, Bharat is also producing Nano DAP today. In the initial years itself, farmers of the country have also started adopting Nano DAP on a large scale.

growth rate. The growth rate of the global market of fertilisers between 2017 to 2022 was 3.8 per cent. During this period, the growth rate of Bharat's fertiliser sector was 9 per cent. The impact of all this is clearly visible on Bharat's agricultural production. Bharat, which was once dependent on other countries for its food needs, is today not only self-sufficient in food grains but is also exporting food grains to other countries on a large scale.

Today, if there is a food crisis anywhere in the world, Bharat contributes in dealing with the situation and the grains produced by the farmers of Bharat reach there. Be it medicines, vaccines, food grains or technology, Bharat is making its mark as a resourceful country. One of the biggest examples of how innovation has been taken to the production and application level in the fertiliser sector is Nano Urea.

Today, Bharat is not only producing Nano Urea on a large scale to meet its own needs but is also exporting Nano Urea to other countries.

It is a matter of pride for all of us that on the lines of Nano Urea, Bharat is also producing Nano DAP today. In the initial years itself, farmers of the country have also started adopting Nano DAP on a large scale.

Now the scientists of our country are doing research on Nano rock phosphate and Nano zinc oxide and their research has reached advanced stage. Similarly, our scientists are also carrying out research and experiments on glauconite nano-particles. Initial experiments have shown that its use has increased the production of biomass. When used on maize crop, it was found that production increased. Similarly, scientists and agricultural research institutions of our country are working on urea which is coated with micronutrients like zinc, etc. Micronutrients play an important role in balanced nutrition of crops. Scientists of our country are working on developing an innovative urea product which is fortified with zinc and other micronutrients. I am confident that in the coming days, this research will also bring revolutionary products and changes in the field of fertilisers.

Be it Nano Urea or Nano DAP, all these successful experiments in the field of fertilisers are the result of Prime Minister Shri Narendra Modi ji's call to promote the use of alternative fertilisers. Potash made from the waste of sugar mills is organic in nature and a good alternative to imported potash, and this has been covered in earlier sections.

Interesting Facts

The annual growth rate of Bharat's fertiliser sector is also more than double the world's annual growth rate. The growth rate of the global market of fertilisers between 2017 to 2022 is 3.8 per cent. During this period, the growth rate of Bharat's fertiliser sector was 9 per cent.

Do You Know?

Now the scientists of our country are doing research on Nano rock phosphate and Nano zinc oxide and their research has reached advanced stage. Similarly, our scientists are also carrying out research and experiments on glauconite nano-particles.

Under GOBARdhan Yojana, making organic fertilisers from agricultural waste and animal waste is also being encouraged. Scientists of our country are also doing research on the enrichment of organic waste. Due to enrichment, microbial activity in organic waste increases and hence it can be used better as fertiliser. Similarly, the Prime Minister has launched Urea Gold in the form of sulphur coated urea in July, 2023. Farmers of our country are also adopting these fertilisers rapidly. Detailed information about these is given in previous chapters of this book.

Scientists of our country have also done research on making organo-mineral fertilisers. Under this, animal waste is mixed with mineral fertilisers. The fertiliser produced by this method is more effective and contains more nutrients, and farmers are getting better results from this type of fertiliser. This type of fertiliser is prepared in the form of granules, powder or pellets. A special feature of this type of fertiliser is that it provides both macro and micro nutrients. The research done so far on this fertiliser has shown that its use has not only resulted in better growth of plants but has also increased the yield. Another feature of this fertiliser is that it can be produced in a decentralized manner in every village. Obviously, such innovative fertiliser production will bring revolutionary changes not only in the fertiliser sector but in the entire agricultural sector.

The Modi government has encouraged the innovation, production and use of bio-capsules as organic fertilisers. The use of bio-capsules is continuously increasing among farmers as compared to traditional fertilisers. Initial studies conducted on the use of bio-capsules are revealing that young farmers are adopting it on a large scale. In recent years, it is also being seen in the agricultural sector of Bharat that educated youths who were working in corporates or running businesses in other sectors are also turning towards agriculture. These people are doing modern agriculture in a very scientific manner. Such youths are using these bio-capsules on a large scale as fertiliser. Encouraging data on the use of bio-capsules by farmers received from Gujarat, Madhya Pradesh, Uttar Pradesh, Kerala, Karnataka, Tamil Nadu, Andhra Pradesh, Telangana, Maharashtra, Himachal Pradesh and other states.

Interesting Facts

Bharat, which was once dependent on other countries for its food needs, is today not only self-sufficient in food grains but is also exporting food grains to other countries on a large scale. Today, if a situation of food crisis arises anywhere in the world, Bharat contributes in dealing with this situation and the food produced by the farmers of Bharat reaches there.

Do You Know?

Be it Nano Urea or Nano DAP, all these successful experiments in the field of fertilisers are the result of Prime Minister Shri Narendra Modi ji's call to promote the use of alternative fertilisers. Potash made from the waste of sugar mills is also included in this. Modi ji's government decided that subsidy will be given on this potash.

Similarly, the work of making alternative fertilisers from seaweed extracted from the sea has also gained momentum in the last nine years. The fertiliser made from seaweed is very useful in terms of soil fertility and plant growth. Fertiliser companies of the country are making different fertiliser products from it. The seaweed extract concentrate produced by IFFCO has been made from red and brown algae. It contains many elements which are essential for the growth of plants.

Similarly, waste from the fishing industry contains a good amount of nutrients like nitrogen, potassium, calcium and magnesium. Work is going on to produce fertiliser from this waste which can be used by farmers who grow fruits and vegetables. It is an absolutely organic fertiliser and is currently being thrown as waste. This provides a great opportunity to convert it into a fertiliser rich in multiple nutrients. Nutrients from the fertilisers made from this are quickly absorbed by the plants. The fish processing industry in Bharat generates 3.5 lakh tonnes of such waste every year. If this large quantity is used in making fertiliser, not only will it provide additional income to the people associated with the fishing industry, but farmers will also be able to get organic fertiliser at a reasonable price.

It is clear from these facts that during the tenure of Prime Minister Shri Narendra Modi ji, innovation work has been done at many levels in the fertiliser sector and this work is going on continuously. Today, the entire country is benefiting from its positive results. I have full faith in the capabilities of our country's scientists and research institutions and I am confident that the benefits of the innovations going on now are going to be more widely available to the farmers in the coming days and through revolutionary changes in the agricultural sector. The "Kisan Samriddhi" (farmers' prosperity) idea of Prime Minister Shri Narendra Modi ji is going to be fully realized in the coming days.

❑ ❑ ❑

We have emphasized on increasing scientific methods and maximum use of technology in the agriculture sector. Today, 22 crore soil health cards have been given to the farmers in the country so that they can get correct information about the health of the soil.

"A lot of work is going on in the country today on biofuels and ethanol to reduce foreign dependence on crude oil and gas. Farmers and farming are directly connected to this work. Vehicles should run on ethanol produced from farmers' produce and bio-CNG should be made from biogas produced from garbage and cow dung—this work is being done today."

19

Crackdown on Black Marketing, Smuggling and Diversion

The major problem of the fertiliser market was that it was a victim of black marketing, smuggling and diversion. Subsidized fertilisers meant for farmers were hoarded and when the demand from farmers increased, black marketing started. Arbitrary prices were collected from the farmers. Another big problem was given the low prices of subsidised fertilisers, these were smuggled to other neighboring countries. One more big problem in the fertiliser sector was that the fertilisers made for agricultural purposes were used illegally in non-agricultural purposes, i.e., in other industries. The Modi government took strict action on all these fronts and today, the positive results are in front of everyone.

To prevent black marketing, smuggling and diversion of fertilisers, the Department of Fertilisers formed "Fertiliser flying squads". These flying squads, in collaboration with State Governments and Union Territories, within a year, raided and inspected 509 units by the end of August, 2023.

To stop black marketing, smuggling and diversion of fertilisers, we started working at many levels. First, the top-20 campaign was started. Under this, information started coming to the Fertiliser Department as to who are the 20 people who buy the most fertilisers at the district level. This helped us to understand, who has legitimate fertiliser needs and who is purchasing excess fertilisers for black marketing. We started collecting data of fertiliser sales at every level and through its analysis we identified the weak links from where there was a possibility of black marketing, smuggling and diversion of fertilisers. Based on these inputs, the department took appropriate steps. The Department of Fertilisers worked with team spirit in collaboration with State Governments and other agencies to prevent black marketing, smuggling and diversion.

Urea used in agricultural purposes was getting diverted and misused in many industries. Among these, plywood industry, animal feed industry and resin industry are prominent. Flying squads of the Fertiliser Department conducted raids and took action against them. By the end of August 2023, 114 units in 10 states and Union Territories were raided and investigated and FIRs were registered against 45 of these units. Action was taken against them as per legal provisions.

Smuggling of fertilisers has been curbed during the tenure of the Modi government. Fertilisers were mainly smuggled from

Interesting Facts

Smuggling of fertilisers was also curbed during the tenure of the Modi government. Fertilisers were mainly smuggled from Bharat to Pakistan, Bangladesh and Nepal. The main reason for this is that the prices of fertilisers in these countries are much higher than in Bharat.

Do You Know?

To prevent black marketing, smuggling and diversion of fertilisers, the Department of Fertilisers working under the Union Ministry of Chemicals and Fertilisers has formed "Fertiliser flying squads". These flying squads, in collaboration with State Governments and Union Territories, raided and inspected 509 units by the end of August, 2023.

Bharat to Pakistan, Bangladesh and Nepal. The main reason for this is that the prices of fertilisers in these countries are much higher than in Bharat. Taking advantage of this situation, smugglers were earning a lot of money by sending fertilisers from Bharat to these countries. But the Government, along with the border states, succeeded in stopping the smuggling of fertilisers across the border. Now neighboring countries are looking for long-term agreements to buy fertilisers from us through official channels. The success of the Modi government's campaigns against smuggling of fertilisers can be gauged from this fact.

Due to well-planned crackdown by the Modi government against hoarding, black marketing, smuggling and diversion, there is no fertiliser crisis in any corner of the country. Earlier, there were many instances of fertiliser shortage in the country. The fertilisers were not rightly reaching the farmers. Farmers had to stand in long queues. Despite this, they were not able to get fertiliser.

With these interventions the problems have been addressed to a great extent. Farmers are getting sufficient quantity of fertiliser and are able to apply fertiliser to their fields as per their need. Because of this, food grain production is increasing and Bharat is making new records every year during the tenure of Prime Minister Shri Narendra Modi ji.

❑ ❑ ❑

In every village, these "Namo Didis" will connect the agricultural system of the country with modern technology. This not only provides them with an additional means of income but also brings a significant change in farming practices.

"In our country, discussions regarding farmers and agricultural policies were limited during the tenure of previous governments. The discourse on empowering farmers was restricted mainly to issues surrounding production and the sale of crops. However, farmers face a variety of challenges in their daily lives. Therefore, our government has made comprehensive efforts to ease every difficulty faced by the farmers."

20

Indian Edge – the first International Magazine

Some international magazines (journals) play a very prominent role in the fertiliser sector all over the world. These magazines have great importance. Especially in giving this information to the whole world about which fertiliser is being sold in which market at what rate. These magazines play a major role in conveying the key information of international markets to the key people in the fertiliser sector.

There was no such magazine in Bharat. In such a situation, the Modi government decided that we have to start an international magazine for the fertiliser sector. Through this magazine, the aim was to provide important information related to the fertiliser sector to the people of the fertiliser sector every month. This magazine was named *Indian Edge*. The Prime Minister of Bharat, Shri Narendra Modi ji released this magazine on October 17, 2022 in New Delhi.

It was decided to add some more content to this magazine as per the needs of the sector. It includes the best practices happening anywhere in the country in the agriculture sector so that farmers from other places can also take inspiration from this and adopt these best practices as per their needs. Apart from this, major decisions taken by the Department of Fertilisers working under the Ministry of Chemicals and Fertilisers of the Central Government and information related to the major policies and schemes of this department were also included in the *Indian Edge* magazine.

Other than information on international markets, detailed information about the status of fertiliser prices in the domestic markets is being made available to the readers through *Indian Edge* magazine. Information about current prices, detailed analysis of price trends in the international and domestic markets is also being given in the magazine, so that companies and various agencies of the government can take appropriate decisions based. Detailed information and analysis about the availability of fertiliser in various markets are also being made available.

Indian Edge is Bharat's first international magazine on fertiliser sector. It covers various information about fertiliser and other related areas. Information about the economics of production of various products in the fertiliser and related sectors is provided in every issue of this magazine so that the stakeholders related to these sectors can take their decisions based on this information. Apart from this, information about

Interesting Facts

Magazines play a major role in conveying the prices of international markets to the key people in the fertiliser sector.

Do You Know?

The Modi government decided that we have to start an international magazine for the fertiliser sector. Through this magazine, the target is to provide important information related to the fertiliser sector to the people of the fertiliser sector every month. This magazine is named *Indian Edge*.

new technology in the fertiliser and other associated sectors are also published in the *Indian Edge* international magazine.

The developments locally and elsewhere in the world are captured so as to establish a line of communication in this era of globalisation.

Apart from this, information on dimensions related to shipping and transportation is given in this magazine, because transportation by sea, rail and road plays an important role at different levels in the fertiliser supply chain. Information related to the strategies of various companies in the fertiliser and related sector and their successful experiments are also given in this magazine.

Indian Edge international magazine publishes authentic analysis on market trends and future along with prices regarding various markets related to fertiliser sector. For example, market analysis related to nitrogen, nitrate and ammonia is included in each issue. A detailed information about various phosphate products and its raw material, i.e., phosphate rock is also disseminated to various stakeholders of the fertiliser sector through this magazine. Similarly, under potash section, global and domestic market information regarding potassium chloride, potassium sulphate and potassium nitrate is also published in this magazine.

Apart from all this, information about sulphur, sulphuric acid, micro nutrients and other important products of fertiliser sector and related sectors is also published in this magazine. With this information, Indian fertiliser companies are able to take decisions based on facts and ensure the supply of fertilisers as per the needs of India's agricultural sector.

This magazine is being published uninterruptedly every month from October, 2022. This magazine has gradually emerged as the leading voice of the fertiliser sector in Bharat. Officials working in the fertiliser sector of the country, key people of fertiliser companies and experts in fertiliser and agriculture wait for every new issue of this magazine. The popularity of *Indian Edge* magazine is increasing with every issue.

This magazine is gaining its own identity even outside Bharat. Fertiliser and agriculture experts from other countries

Interesting Facts

Indian Edge **magazine also includes major decisions taken by the Department of Fertilisers working under the Ministry of Chemicals and Fertilisers of the Central Government and information related to the major policies and schemes of this department.**

are also taking interest in *Indian Edge* magazine. The team running this magazine is confident that the international fame of this magazine will increase further in the coming days.

❑ ❑ ❑

APPENDICES

1

Bharat's New Initiatives for Sustainable Fertiliser Management

– Sh. Narendra Singh Tomar
Speaker, Madhya Pradesh Legislative Assembly and
Former Union Minister of Agriculture and Farmers Welfare

Sustainable agriculture is essential in ensuring that Bharat achieves its national security goal for 1.4 billion Bharatiya. Under the guidance of Shri Narendra Modi ji, the Ministry of Chemicals and Fertilisers took proactive initiatives to address the growing issue of imbalanced chemical fertiliser usage. These initiatives aim to structurally transform Bharatiya Agriculture through synergizing various efforts of policy intervention, investment, financial support, technological intervention and value addition. Recognizing the situation's urgency, on June 28, 2023, the Cabinet Committee on Economic Affairs (CCEA) approved a series of initiatives to continue the Urea Subsidy Scheme and promote the adoption of organic fertilisers, with a total outlay of Rs 370,128.7 crore. This reflects the government's strong commitment to sustainable agriculture and the well-being of more than 12 crore farmers having 141 million hectare of land holding. The major initiatives taken by the government are as under:

Extending the Urea Subsidy Scheme:
The CCEA has granted an extension to the Urea Subsidy Scheme till March 31, 2025, with the allocation of Rs 368,676.70 crore. This extension encompasses the financial years from 2022-23 to 2024-25. With the thrust of Modi government towards indigenous

production, the country has increased the urea production capacity of the nation from 207.54 LMT in 2014-15 to 283.74 LMT in 2022-23. This enhanced production with urea subsidy extension will ensure the accessibility of affordable urea to farmers throughout the nation.

Strengthening of Nano Urea Ecosytem:
Bharat has developed the world's first Nano Urea Liquid indigenously and produced it commercially to empower Bharatiya farmers with an innovative, eco-friendly and economical product. Till March 2023, 7.65 crore bottles (equivalent to 33.6 LMT of conventional urea) have been produced, and 5.42 crore bottles have been sold. By 2025-26, eight Nano Urea plants with a production capacity of 44 crore bottles equaling 195 LMT of conventional urea will be commissioned. Farmers are also introduced to Nano DAP as an effective and cost efficient alternative for conventional DAP.

Under the aegis of Aatmanirbhar Bharat, the government has set up and revived six urea production units viz. Chambal Fertilisers and Chemicals Ltd. - Kota Rajasthan, Matix Ltd. Panagarh West Bengal, Ramagundam-Telangana, Gorakhpur-UP, Sindri-Jharkhand, and Barauni-Bihar. These indigenous production units and Nano Urea plants will reduce current import dependency on urea and finally make us Aatmnirbhar in urea by 2025-26.

Interesting Facts

CCEA has approved extension of Urea Subsidy Scheme till March 31, 2025 with an allocation of Rs 368,676.70 crore. This extension covers the financial years from 2022-23 to 2024-25.

Do You Know?

Under the Market Development Assistance (MDA) component, the government will provide assistance at the rate of Rs 1,500 per metric tonne for organic fertilisers produced in plants linked to the unique, multi-faceted and waste-to-wealth GOBARdhan initiative.

Supporting Organic Fertilisers through GOBARdhan:
Under the Market Development Assistance (MDA) component, the government will provide support at the rate of Rs 1,500 per metric ton for organic fertilisers produced at plants associated with the unique, multi-pronged, waste-to-wealth GOBARdhan initiative. This holistic and integrated approach incorporates various biogas and renewable energy schemes, waste management programs, and sanitation initiatives. With a total outlay of Rs 1,451.84 crore for FY 2023-24 to FY 2025-26, this funding will also include research gap funding of Rs 360 crore, further bolstering the growth of organic fertilisers in Bharat. To provide farm inputs (including organic fertiliser) and services as a one-stop solution for farmers and empower them, about one lakh unique model farm inputs and service outlets, “Pradhan Mantri Kisan Samriddhi Kendras”, have been established nationwide.

Introducing PM-PRANAM:
As an eco-friendly and sustainable initiative, the government introduced the PM Programme for Restoration, Awareness Generation, Nourishment and Amelioration of Mother-Earth (PM-PRANAM). This scheme incentivizes states and Union

Territories to embrace natural farming practices, promote alternative fertilisers and encourage the balanced utilization of chemical fertilisers. The incentive funds allocated for 2025-26 will be disbursed in 2026-27, highlighting the government's long-term vision for sustainable agriculture.

Innovative Urea Gold:
The approval also introduces a new-age, value added urea called "Urea Gold," featuring sulphur coating. This innovative coating will make available an important secondary plant nutrient sulphur to crops. "Urea Gold" reduces urea consumption and enhances crop productivity by ensuring slow release and improved nitrogen use efficiency. This sustainable fertiliser option aligns with the government's efforts to improve nutrient management and optimize urea manufacturing processes.

The recent approvals by the Cabinet Committee on Economic Affairs demonstrate the unwavering commitment of the Bharatiya government to sustainable agriculture and the well-being of farmers. The extension of the Urea Subsidy Scheme, the introduction of PM-PRANAM, support for organic fertilisers through GOBARdhan and the innovative Urea Gold contribute to building a more eco-friendly and efficient agricultural sector. In addition to these initiatives, the government has also launched mass awareness

Interesting Facts

With the Modi government's emphasis on indigenous production, the country has increased its urea production capacity from 207.54 lakh metric tonnes (LMT) in 2014-15 to 283.74 LMT in 2022-23.

Do You Know?

The government has also launched a large-scale public awareness campaign to promote balanced use of chemical fertilisers and adoption of alternative fertilisers like organic, bio and nano fertilisers. The goal of these efforts is to improve the soil and nourish Mother Earth.

campaigns to promote the balanced use of chemical fertilisers and adopt alternate fertilisers such as organic bio and nano fertilisers. These efforts are aimed at ameliorating the soil and nourishing Mother Earth.

As the Modi government marks nine years of dedicated service to the nation, it is crucial to acknowledge the multitude of welfare schemes implemented to empower and uplift the agriculture sector. The government envisions a new Bharat guided by the principle of "Sabka Saath, Sabka Vikas, Sabka Vishwas," with the welfare of farmers being an integral aspect of this vision. These schemes aim to enhance farmer's livelihoods, promote sustainable agricultural practices, and foster overall development in the agricultural sector. It underscores the government's commitment to creating a prosperous and inclusive Bharat, where farmers play a pivotal role in the nation's progress.

(The article was published in Business Standard on July 14, 2023)

❑ ❑ ❑

2

An Aatmanirbhar Move

– Dr. Mansukh Mandaviya

"When the world is in crisis, we must pledge—a pledge that is greater than the crisis itself. We must endeavour to make the 21st century, Bharat's century and the path to achieving this is self-reliance."

This powerful statement by Prime Minister Narendra Modi encapsulates the essence of Modinomics in more ways than one. His government has just concluded four years after being re-elected to office in 2019. It is time to revisit these four years of Modi 2.0. They have been extraordinary for the sheer number of opportunities seized to establish new milestones. PM Modi's exceptional management of Covid-19 crisis and the ensuing challenges conveyed the message to concentrate, recalibrate and advance without excessive indulgence. This is embodied in the plan for a self-reliant Bharat, Aatmanirbhar Bharat.

PM Modi's Vocal for Local initiative is not just about import substitution; it is about creating a self-sustaining and self-generating economy. Bharat's department of fertilisers has made significant strides in achieving self-reliance in the sector. The government has shown foresightedness in securing fertiliser supplies during the crisis precipitated by the Russia–Ukraine war. Despite facing challenges such as scarcity of raw materials, gas, oil, rock phosphate and potash, the government has forged long-term agreements and created joint ventures with resource-rich nations.

Industry has been encouraged to set up overseas joint ventures to secure raw materials. Joint venture plants have been set up in some countries with buyback agreements and assured offtake agreements for supply of 10 LMT of rock phosphate and 6.5 LMT of phosphoric acid. As a result of strategic partnership with countries such as Jordan, Saudi Arabia, Oman, Canada, Russia, Morocco, Israel, Senegal, Tunisia and South Africa, India has secured a supply of 157 LMT for various fertilisers for three years and 32 LMT for four years.

The government has also encouraged the domestic industry and public sector undertakings to sign long-term agreements for annual import of raw materials/intermediates such as ammonia, phosphoric acid and sulphur. These initiatives have enabled Bharat to diversify its product portfolio and produce more NPK complex, which offers not only wider rage of products to farmers but also ensures balanced nutrition.

There has been a remarkable improvement in the opening stocks of DAP and MOP fertilisers in Kharif 2023. The opening stocks have increased from 7.73 LMT, 2.47 LMT and 13.8 LMT

Interesting Facts

The Government has also encouraged the domestic industry and public sector undertakings to sign long term agreements for import of raw materials/intermediates such as ammonia, phosphoric acid and sulphur.

Do You Know?

The Government has shown foresight in ensuring fertiliser supplies during the crisis arising out of the Russia-Ukraine war. Despite facing challenges such as shortage of raw materials, gas, oil, rock phosphate and potash, the government has entered into long-term agreements and set up joint ventures with resource-rich countries.

in Kharif 2022 to 27 LMT, 14 LMT and 13.2 LMT respectively.

The Department of Fertilisers has positioned Bharat as a key player in the global fertiliser supply chain.

Bharat's bold move towards self-reliance in fertiliser is a testament to the government's commitment to ensuring food security for its citizens and fulfilling PM Modi's vision of Aatmanirbhar Bharat.

(The article was published in The Indian Express on June 13, 2023)

❑ ❑ ❑

3

PMKSK Can Transform Bharat's Farming Sector

– Dr. Mansukh Mandaviya

Around 65 per cent of the country's population lives in rural areas and 47 per cent is dependent on agriculture for livelihood, according to the Economic Survey 2022-23. Agriculture is a time-bound activity that requires the right agri-inputs at the right time to maximise production and productivity. Agri-inputs are essential ingredients of agriculture and an efficient delivery system plays a crucial role in the growth of farm income.

In Bharat, the network of input services is fragmented and there are separate dealer networks for seeds, fertilisers, pesticides and implements functioning in silos. In addition, facilities for testing of soil, seeds, fertilisers and information on different schemes reach farmers through different agencies in a disjointed and piecemeal manner. This approach fails to provide technology-based holistic information for taking informed decisions. Moreover, farmers spend their precious time procuring agri-inputs. The solution lies in providing holistic solutions to the farmers under one umbrella, backed by the government, that the farmer can trust and follow. In the past, the private sector has tried to replicate the model of umbrella centres for farmers' inputs and services, but was largely unsuccessful.

Thus, the idea of transforming the existing fertiliser retail shops into Pradhan Mantri Kisan Samriddhi Kendra (PMKSK), a one-stop shop solution, has come up as a sustainable and long-term solution to cater to the needs of the farmers. This programme has ushered in a revolutionary transformation in the country's agricultural sector. With a vision to empower farmers and boost agricultural growth, this initiative aims to simplify access to agricultural inputs, information and services, uplifting the lives of farmers and contributing to the nation's progress.

Under the PMKSK initiative, around 2.5 lakh active retail fertiliser shops are undergoing a phased conversion into comprehensive one-stop shops for farmers. At its core, PMKSK focuses on offering a diverse range of agri-inputs, including fertilisers, seeds, pesticides and small farm machinery to the farmers, including drone services, in a hassle-free manner. A vital aspect is the provision of soil and seed testing facilities. By equipping farmers with knowledge about their specific soil and crop conditions, these facilities enable informed decision-making, optimised resource utilisation and higher yields. This promotes precision agriculture and resource-efficient farming practices. Moreover, PMKSKs serve as a knowledge hub, disseminating crucial information on crops and government welfare schemes. Bridging the information gap, these centres empower farmers to make informed choices, thereby enhancing their efficiency and income.

The aim is to simplify the lives of millions of farmers. Recent developments demonstrate its success. In a significant event

Interesting Facts

Under the PMKSK initiative, around 2.5 lakh active retail fertiliser shops are being converted into comprehensive one-stop shops for farmers in a phased manner.

Do You Know?

The idea of converting existing fertiliser retail shops into Pradhan Mantri Kisan Samriddhi Kendra (PMKSK), a one-stop shop solution, has emerged as a sustainable and long-term solution to meet the needs of farmers. This programme has initiated revolutionary changes in the agriculture sector of the country.

named PM-Kisan Sammelan held in Sikar, Rajasthan, on July 27, the Prime Minister dedicated 125,000 PMKSKs to the nation. The widespread participation of nearly 20 million farmers across the country shows an affirmative stamp by the end-users on this initiative.

The success can also be gauged from the feeling of oneness and a sense of pride among different stakeholders. A recent survey indicated that there is already a 15-20 per cent increase in the number of farmers visiting the centres and it is found that they are happy with the ambience of PMKSK and the inputs available. The sale of Nano Urea too has reached Rs 6 crore through these PMKSKs. Drone entrepreneurs supported by fertiliser companies are slowly associating with PMKSKs for promoting the spraying of fertilisers and chemicals in a better way. The increased bonding between states, KVKs and dealers has resulted in the successful dissemination of knowledge and services.

The PMKSK initiative is proving to be a game changer in empowering the farmers and strengthening the backbone of the nation's economy. By simplifying access to agricultural inputs and knowledge, empowering farmers with modern technologies and fostering sustainable practices, PMKSKs are ushering in a new era of prosperity for the farming community. With unwavering government support and collective efforts from stakeholders, PMKSKs will continue to drive positive change, strengthening the agricultural backbone of the nation and contributing to a thriving and self-reliant agrarian ecosystem.

(The article was published in Hindustan Times on August 11, 2023)

❑ ❑ ❑

4

"PMKSK Will Empower Farmers"

Edited excerpts of the address of Dr. Mansukh Mandaviya, Union Minister of Chemicals and Fertilisers on the occasion of the launch of "One Nation, One Fertiliser" by Prime Minister Shri Narendra Modi in New Delhi on October 17, 2022.

Respected Prime Minister Shri Narendra Modi Ji, Union Agriculture and Farmers Welfare Minister Shri Narendra Singh Tomar Ji, Agriculture Ministers of the states joining from different places, agriculture experts and my farmer brothers & sisters, today I welcome you all on this occasion.

I am proud to welcome Prime Minister Shri Narendra Modi ji on this occasion. I am feeling proud because today our Prime Minister is releasing the 12th installment of the Pradhan Mantri Kisan Samman Nidhi, simultaneously inaugurating 600 Prime Minister Kisan Samriddhi Kendras, launching "One Nation, One Fertiliser" scheme under the Pradhan Mantri Bhartiya Jan Urvarak Pariyojna and releasing *Indian Edge*, an international magazine dedicated to fertiliser sector.

Respected Prime Minister, your participation in this programme is a symbol of your commitment towards the farmers of the country. This shows how, under your leadership, new initiatives are continuously being taken to improve the lives of farmers. Honourable Prime Minister Sir, when you took over the reins of the country, you had said that this government is a government of farmers and that this government is a government of the poor. During your tenure of more than eight years, many such steps have been taken which have strengthened and empowered the farmers of the country.

Honourable Prime Minister Sir, you have always worked with a total approach in agriculture sector. Be it adopting new technology in agriculture, promoting the use of smart technology or providing better markets to farmers' produce, a lot of work has been done on all these fronts under your leadership. Therefore, on one hand, Bharat's agricultural production has reached to record levels, while on the other hand, farmers are getting better and fair prices for their produce.

Prime Minister Sir, during your tenure, the agriculture research has got a massive boost. For this reason, Bharat became the first country in the world to start commercial production of Nano Urea. Today, farmers of the country are using Nano Urea on a large scale. On the other hand, on your appeal, the inclination of farmers towards natural farming is also increasing in the entire country. The level of awareness among farmers regarding balanced use of chemical fertilisers has also increased continuously.

Respected Prime Minister Sir, with your efforts, the 600 Pradhan Mantri Kisan Samriddhi Kendras that are being

Interesting Facts

Bharat became the first country in the world to start commercial production of Nano Urea. Today, farmers of the country are using Nano Urea on a large scale.

Do You Know?

Now with only "Bharat" brand name, farmers will get all types of fertilisers across the country. Under this project, farmers will get fertilisers across the country in "single bag design" under the names of "Bharat Urea", "Bharat DAP", "Bharat MOP" and "Bharat NPK".

inaugurated today will empower the farmers in many ways. On one hand, this will work as a model fertiliser retail shop, while on the other hand, different agro inputs will also be made available to the farmers. Apart from this, information about government schemes, application process for these schemes, good practices of crops, information about prices of different markets, soil testing facilities, etc. will also be made available to the farmers here under one roof. These centres will work as a one-stop shop for farmers for information and facilities related to farming.

Prime Minister Sir, today under the Pradhan Mantri Bhartiya Jan Urvarak Pariyojna, you are launching "One Nation, One Fertiliser". Now, farmers will get all types of fertilisers in the entire country with only the single brand name "Bharat". Under this project, farmers will get fertilisers across the country in "single bag design" under the names of "Bharat Urea", "Bharat DAP", "Bharat MOP" and "Bharat NPK". Due to different brand names, there is confusion among the farmers. They think that perhaps one brand of fertiliser is better than others. However, it is not like that. Due to the same brand name, there will be no confusion and farmers will find it convenient to buy fertilisers.

Prime Minister Sir, *Indian Edge* magazine will provide new and useful information about the fertiliser sector to the companies of this sector and the people interested in the fertiliser sector. Besides this, there will also be success stories of agriculture and fertiliser sectors in the magazine.

Respected Prime Minister Sir, under your leadership, all the steps being taken today are aimed at empowering the farmers, strengthening the agriculture sector and increasing agricultural production. Prime Minister Sir, on this occasion, I once again welcome you and express my gratitude to you.

Thank you very much.

❑ ❑ ❑

5

"Bharat Will Soon Achieve Self-sufficiency in Urea"

Dr. Mansukh Mandaviya, Union Minister of Chemicals and Fertilisers, in conversation with Himanshu Shekhar, answered questions about the complaints of urea shortage coming from some parts of the country, its current situation and various immediate and long-term steps the government is taking to solve this problem. Here is the edited excerpt.

This year again there are reports of urea shortage in Kharif season from some parts of the country. How is the Union Government looking at this situation?

We never have any shortage of stock. The Union Government already seeks month-wise estimates of the total urea requirements for the season from the states. When urea reaches at the rake point in any state, the responsibility for district wise allocation and distribution rests with the concerned state. It has been observed on some occasions that some problems arise in district wise allocation and distribution within the states. Because of this, the reports of artificial urea shortage is coming from some places.

How is the Central Government monitoring the demand and supply situation and what are your plans to deal with the situation of increased demand in any particular season?

The Department of Fertilisers has created a dashboard to ensure adequate supply as per the demand at every location. On this dashboard, information about fertiliser sales, availability and requirement up to the district level is available on daily basis. If the states use this dashboard properly and take allocation and distribution decisions based on the information available on it, then no issue of artificial urea shortage will arise.

Should the farmers of the country be assured that there will be no shortage of urea in the current Kharif season and the coming Rabi season?

We have made arrangements for the amount of urea required for the Kharif season as well as the Rabi season. Domestic production is continuously increasing and we make preparations in advance to import the required quantity of urea. By December 2022, there

> Interesting Facts
>
> **The Department of Fertilisers has created a dashboard to ensure adequate supply as per the demand at every location. On this, dashboard information about fertiliser sales, availability and requirement up to the district level is available on daily basis. If the states use this dashboard properly and take allocation and distribution decisions based on the information available on it, then no issue of artificial urea shortage will arise.**

Do You Know?

Prime Minister Narendra Modi dreams of creating a self-reliant Bharat (Aatmanirbhar Bharat). The government is committed to make the country self-reliant in urea. The way our conventional urea and Nano Urea plants are being started, we hope that soon Bharat will not have to import urea and we will be able to produce more than the demand. The government is also emphasizing on increasing the use of alternative fertilisers.

will be availability of approximately 275 lakh MT of urea including domestic production and imports. This is slightly more than our demand till December. Therefore, farmers will not have to face any kind of urea shortage.

When information about urea shortage reaches the Central Government from somewhere, what does the Government do for its immediate solution?

The Central Government holds weekly review meetings with the states. The directors of the agriculture departments of the states participate in such meetings. Through this meeting we remain updated about the demand and availability. We also do a weekly review with the fertiliser companies. This makes the actual situation of production and supply clear to us. If any problem comes to light in both these review meetings, the Central Government immediately takes necessary steps to solve it.

What long-term measures is the Government taking to solve the problem of urea shortage?

The Central Government is working towards a permanent solution. Prime Minister Narendra Modi has dreamed of creating a self-reliant Bharat (Aatmanirbhar Bharat). The Government is committed to make the country self-reliant in urea. The way our conventional urea and Nano Urea plants are being commissioned, we hope that soon Bharat will not have to import urea and we will be able to produce more than our demand. The Government is also emphasising on increasing the use of alternative fertilisers. With the use of alternative fertilisers, not only will our agriculture and farmers become more prosperous, but the health of our Mother Earth will also improve and agricultural products will also be of better quality.

The Central Government is continuously motivating farmers to increase the use of Nano Urea. What is its role in solving the problem of urea shortage?

Bharat is the first country in the world to commercially produce Nano Urea. Currently, Nano Urea is being produced from the IFFCO plant in Kalol, Gujarat. Seven more plants are going to start in the coming days. When urea has to be sprayed on the crop for the second time after sowing, then the use of Nano Urea can be increased instead of conventional urea. On one hand, this increases crop productivity and fertiliser use efficiency, while on the other hand, farmers also get financial savings. Besides, it is also environment friendly. It is also useful in reducing water pollution and maintaining good soil health. Nano Urea is also useful in reducing the production cost of farmers and increasing their income.

(This interview was published in India Today Magazine on September 11, 2022)

❑ ❑ ❑

6

"Kisan Samriddhi Kendra Will Work to Change the Future of Farmers"

Edited excerpts of the special interview of Dr. Mansukh Mandaviya, Union Minister of Chemicals and Fertilisers, regarding Pradhan Mantri Kisan Samriddhi Kendra, telecast on DD News on August 4, 2023.

What is the thinking behind Pradhan Mantri Kisan Samriddhi Kendra and what benefits will farmers get from it?

First of all, I would like to tell you about the Kisan Samriddhi Kendras. These are developed as a result of the thinking of Hon'ble Primc Minister Modi ji that there should be a Kendra which shall act as a "one-stop solution" for farmers to provide necessary inputs and act as a medium to have direct communication with them. Kisan Samriddhi Kendra aims to achieve the objective of providing a platform to educate the farmers so that they can avail the services of the Government of Bharat.

What types of services are covered in these centres?

It is noticed that there was no primary medium to convey the benefits of the schemes of the Government of Bharat to the farmers directly in their vicinity. The menace of middlemen is hurting the delivery of goods and services to our farmers. The menace is so huge that the farmers are forced to pay these middlemen for the services which are ideally free for them. In these circumstances, Kisan Samdriddhi Kendras have emerged as a breakthrough solution to these problems. It is a platform at district, block and village level having AV facilities-backed with Internet, acting as one stop solution for various services such as information dissemination, soil testing facilities, various inputs such as fertilisers, seeds, agricultural equipment and related farm inputs. In fact, Kisan Samriddhi Kendras will work as a catalyst to transform the future of farmers. The government has started "One Nation, One Fertiliser". All types of fertilisers, be it chemical or organic, urea, DAP, NPK, all are available at Kisan Samriddhi Kendras.

Earlier, every company had built their own brand and farmers of the particular region were getting used to it. Chemically, urea is the same product having 46 per cent nitrogen across all brands. Similarly, NPK fertilisers including DAP have similar nutrient content irrespective of their brands. The branding has no significance in the fertiliser segment. After a brand becomes popular, farmers think that this fertiliser is good and it creates a demand for that particular fertiliser in that particular region. In case of unavailability of that particular brand in their pre-dominant area, an artificial shortage is created, despite of availability of

Interesting Facts

We have developed nano fertilisers and Urea Gold which are also available at Kisan Samriddhi Kendras. It is evident that multiple benefits to farmers will be available through "One Nation, One Fertiliser".

Do You Know?

Under one roof, farmers will get farm equipment, seeds, testing facility and fertilisers, and farmers will also get redressal of their complaints.

other brands of fertiliser containing similar nutrient content. With "One Nation One Fertiliser" in force, we only have a single brand, i.e., "Bharat". Different fertilisers will be available as Bharat Urea, Bharat NPK, Bharat DAP, etc. This will also help in stopping the criss-cross movement of fertilisers.

How are "Kisan Samriddhi Kendras" making the lives of farmers easier as "one-stop centres"?

We had dedicated 600 Kisan Samriddhi Kendras to the country in October, 2022. During the past 6-7 months, these Kendras have delivered good experience to farmers. While visiting these Kendras, farmers get to know about scientific agricultural practices. If a farmer faces a problem and needs technical advice, he can avail the facility of teleconsultation. Through this facility, he can directly connect with scientists of agricultural colleges and universities of that area within no time and get solution to their issues immediately. Based on our initial experience of these Kendras, we thought to take it further. This is an excellent platform to put forth the views of the Government for the benefits of the framers.

Before this initiative, there was only one Krishi Vigyan Kendra in every district. Now there have been 1.25 lakh Kisan Samriddhi Kendras across the nation. There are about 10-50 Kisan Samriddhi Kendras in every district. These kendras will help the Krishi Vigyan Kendras in servicing the farmers and will create an ecosystem for disseminating useful information. For example, if some important information is to be communicated to the farmers before the start of a crop season, we can connect through 1.25 lakh Kisan Samriddhi Kendras. When these centers were launched, about 2 crore farmers have participated in that programme, highlighting its outreach.

Imagine the scenario where 2 crore farmers can directly be trained using this platform. On this medium myself, Union Agriculture Minister, along with various functionaries can put forth their views effectively. The farmers can communicate amongst themselves about better agricultural practices. The huge benefits of breakthrough products such as Urea Gold can be disseminated widely. The Kisan Samriddhi Kendra is a unique platform for transforming lives of farmers.

The Government is promoting spraying of fertilisers through drones. Please comment.

It's a good programme and it has taken off impressively. This facility is available at our Kisan Samriddhi Kendras where drones are now available on rent. Currently, there are about 1,200-1,300 drones available at Kisan Samriddhi Kendras across the country. These flying machines are revolutionizing the agriculture as nano fertilisers such as Nano Urea and Nano DAP and various water

Interesting Facts

The Prime Minister has created the Kisan Samriddhi Kendras with the objective of providing a platform to educate the farmers by easily communicating with them, and to help them avail the services of the Government of Bharat.

Do You Know?

If the farmers have to fill the form to avail the benefit of any government assistance scheme, they will get the facility to do so. From one point of view, Kisan Samriddhi Kendras will work as a catalyst to transform the future of farmers.

soluble fertilisers and pesticides can be sprayed efficiently using drones. It not only saves time and labour but is economical too. These products can easily be sprayed together over an area of 20–25 acres in a very short span of time. Farmers are getting increasingly curious about drones and its use is increasing now. It is the vision of our Hon'ble Prime Minister which has motivated us to use such technology for increasing production and income of farmers. Our Kisan Samridhhi Kendras will play a pivotal role in promoting the usage of drone widely.

What is the difference between different levels of Pradhan Mantri Kisan Samriddhi Kendras?

Under this scheme, we have decided that all the PACS in the country, including FPO and regular fertiliser outlets, have to be converted into Kisan Samriddhi Kendras. For that, we have decided to make various facilities available at Kisan Samriddhi Kendra based on their location. For example, in the village level PMKSKs, if we cannot have elaborate soil testing facilities, at least we can keep a drone. We can make quality seeds and various fertilisers including Nano Urea and Nano DAP available. In addition to these, more facilities can be made available at block-level Kisan Samriddhi Kendras. Moreover, in case of district level PMKSKs, in addition to the facilities available at block level, some other facilities including farmers/retailers training center are being provided.

Although this is an initiative of the Government of Bharat, it should equally be used by state governments as well for the benefit of the farmers. If the Honourable Chief Minister of the state feels that he wants to give a message to the farmers, then he can use this platform. This platform can similarly be used by other authorities including Agriculture Ministers of the state.

How will PM-PRANAM Yojana prove to be helpful in achieving the goal of self-reliance in the fertiliser sector?

In the fertiliser sector, various raw materials are required for production of various fertilisers. For example, we do not have fertiliser minerals used to make MOP. We have to import it. In spite of these challenges, we are strategizing to move towards self-reliance in many segments of fertilisers. We have now achieved indigenous production capacity to annually produce about 285 LMTs of urea. Our consumption however is 350 lakh metric tons. So, even today, we need to import 60–70 LMTs annually. During the coming years, we have planned to promote the use of Nano Urea and Sulphur coated urea, i.e., Urea Gold. Also, we will promote natural farming in a big way. Under this PM-PRANAM Yojana, if a state reduces the consumption of chemical fertilisers and switch to alternative

Interesting Facts

These flying machines are revolutionizing the agriculture as Nano fertilisers such as Nano Urea and Nano DAP and various water soluble fertilisers and pesticides can be sprayed efficiently using drones. It not only saves time and labour but are economical too.

Do You Know?

As we are moving towards alternative fertilisers, to have effective adoption at a faster pace, we need to create awareness. In order to achieve this objective, the Centre and the states should work together through "PM-PRANAM" initiative. Hence, a three-tier system of Kisan Samriddhi Kendras has been created for the larger benefit of our farmers.

fertilisers, we will pass the part of subsidy saved to that respective state as monetary benefit for promoting the use of these alternative forms. This will encourage the state to use alternative fertilisers.

Why is the Government putting so much emphasis on promoting the use of alternative fertilisers?

As per various studies and research across the world, excessive use of chemical fertilisers and pesticides increases the risk of diseases like cancer. While ensuring the food security of our nation by maintaining production, the use of chemical fertilisers should also be reduced. As, we are moving towards alternative fertilisers to have effective adoption at a faster pace, we need to create awareness. In order to achieve this objective, the Centre and the states should work together through PM-PRANAM initiative. In this scheme of PM-PRANAM, through encouragement by Central Government, we aim to have proactive participation of the states for promotion of use of alternative fertilisers. States should be encouraged to save subsidies and motivate farmers to adopt alternative fertilisers. We proposed to pass monetary benefit in lieu of subsidy thus saved.

With this financial support, concerned states can, in turn, use other schemes in the states for promoting alternative fertilisers such as Nano Urea and Urea Gold.

❑ ❑ ❑

7

Innovative Nano-fertilisers for Eco-friendly Green Agriculture

– Dr. Mansukh Mandaviya

Fertiliser remains one of the critical inputs for increasing food production for providing sustenance to 1.4 billion population of our country. In the last 9 years, our efforts have resulted in enhanced production capacity of urea, a major fertiliser for nitrogen (N) source, which has gone upto 283.74 LMT/ year, a quantum jump from 207.54 LMT/ year in 2013-14. Similarly, the fertiliser industry (including public, cooperative and private players) has been also actively promoting other major nutrients of phosphate (P) and potash (K)-based fertilisers. Balanced use of NPK fertilisers is must for sustainable agriculture. However, in the quest for enhancing crop productivity and to feed increasing population on continuum basis, the vital aspect of soil fertility, soil and water conservation are getting compromised due to excess and imbalanced use of fertilisers. As a result soils have become hungry for nutrients and thirsty for water which is exhibited in their fertility status. This led to requirement of an innovative technology for increasing Fertiliser Use Efficiency (FUE) and to save soil and environment. The new age nano technology has come as an answer to solve this problem and for producing more from less, i.e., to increase input use efficiency with less environment footprints.

Taking inspiration from clarion call of our Honourable Prime Minister on Make in India, Indian scientists and engineers have developed Nano Urea (liquid), the first indigenously developed nano fertiliser. Nano Urea is notified under Fertiliser Control Order (FCO) in 2021 after vigorous field trials and passing strict bioefficacy & biosafety testing protocols as per Department of Biotechnology's "Guidelines on nano agri-inputs". It is a perfect example of initiative undertaken under the "Aatmanirbhar Krishi" and "Aatmanirbhar Bharat" during the Amrit Kaal for ensuring food and nutritional security of the nation in perpetuity. As of now, Nano Urea plants at Kalol in Gujarat, Phulpur and Aonla in Uttar Pradesh with capacity of 17 crore bottles have been set up. By 2025, other nano plants at Nangal, Trombay, Bengaluru, Deogarh, Guwahati and several other places would be ready that will increase Nano Urea production capacity to more than 44 crore bottles/year which will be equal to more than 195 LMT of conventional urea.

Nano Urea costs 16 per cent less than a bag of urea and can be carried easily by farmers to their fields, a fact appreciated by majority of young farmers that has translated into increased sale of Nano Urea bottles. Nano Urea has benefit in term of logistics and storage cost too. A conventional fertiliser rake (BCNHL) carries 69,600 bags of conventional urea, while it can carry 29 lakh bottles of Nano Urea bottles correspondingly. Similarly, one 24 MT truck carrying

Interesting Facts

Nano DAP is a recently introduced nano fertiliser. This is another achievement in terms of crop nutrition. Further research towards development of Nano NPK, Nano Zinc, Nano Copper, Nano Boron, Nano Sulphur, etc., in the field of fertilisers will definitely promote eco-friendly sustainable agriculture in future.

Do You Know?

Taking inspiration from our Honourable Prime Minister's call about "Make in India", Bharatiya scientists and engineers have developed Nano Urea (liquid), the first indigenously developed nano fertiliser.

533 bags of urea can carry 30,000 bottles of Nano Urea. A Nano Urea plant has advantage over a conventional urea plant in terms of capex, energy consumption and less carbon dioxide emission. Nano Urea (liquid) exhibits high nutrient use efficiency (NUE) and its adoption results in enhanced agricultural productivity, quality of produce, farmers profitability and reduction in soil, water and air pollution, logistics and warehousing cost. The field experiments of ICAR in Kharif 2021 with Nano Urea at 20 locations on rice, maize, finger millet, pearl millet and ginger indicated yield advantage of 3-8 per cent and savings of 25-50 per cent for the farmers.

In the last 3 seasons, 6.5 crore bottles (500 ml each) of Nano Urea Liquid have been consumed by 192 lakh farmers as a possible alternate for foliar dose of urea covering an area of 150 lakh hectares. There are very encouraging results at farmers' field level too with Nano Urea usage. In comparison to 2021-22 during 2022-23, Nano Urea sales increased by 55 per cent whereas conventional urea registered marginal growth of 4 per cent. In 2022-23, decrease in consumption of conventional urea is observed in states of Punjab, Kerela, Tamil Nadu, Jammu & Kashmir, Himachal Pradesh, Uttarakhand, Jharkhand, and Assam whereas correspondingly, there has been increase in the consumption of

Nano Urea in these states, for instance, consumption of Nano Urea has increased by 27 per cent in Punjab, 20 per cent in Kerala, 67 per cent in Uttarakhand, 49.6 per cent in Jammu & Kashmir in 2022-23 compared to 2021-22. Similarly, sales of Nano Urea increased in Kharif 2023 over last Kharif, 2022 whereas NCU sale reduced in the states of Himachal Pradesh, Jammu & Kashmir. A total of 189 districts showed reduction in conventional urea consumption during the year 2022-23 compared to 2021-22; out of these, 130 districts in 16 states registered average 12 per cent reduction in conventional urea consumption as well as 76 per cent increase in Nano Urea consumption. In Kharif 2023, reduction in conventional urea sales due to use of Nano Urea was observed in 113 districts in comparison to Kharif 2022.

Nano DAP, the recently included nano fertiliser, is another a feather in the cap of crop nutrition that can reduce the consumption of conventional DAP which eventually will lead to decreased imports. Further research happening in fertiliser sector towards development of Nano NPK, Nano Zinc, Nano Copper, Nano Boron, Nano sulphur, etc., will surely lead to ecofriendly sustainable agriculture in future.

Interesting Facts

So far, Nano Urea plants with a capacity of 17 crore bottles have been set up at Kalol in Gujarat and Phulpur and Aonla in Uttar Pradesh. By 2025, other nano plants will be ready at many places including Nangal, Trombay, Bengaluru, Deogarh and Guwahati, taking the Nano Urea production capacity to more than 44 crore bottles per year.

Do You Know?

During the last three seasons, 192 lakh farmers have used 6.5 crore bottles (500 ml each) of Nano Urea Liquid as an alternative to conventional urea on 150 lakh hectares of land. Even at the field level of farmers, the use of Nano Urea is showing very encouraging results.

The use of nano fertilisers will also help in furthering the objective of PM Programme for Restoration, Awareness, Nourishment and Amelioration of Mother Earth (PM-PRANAM) launched to promote alternate fertilisers and balanced use of chemical fertilisers. Accrued benefits in terms of reduction in Green House Gas (GHGs) emission, saving of people's subsidy, transportation and other sundry costs would be manifold that would ultimately contribute in the nation building.

Further, massive awareness campaigns and farmer connect programmes have been taken up leveraging the advantage of existing institutional structures like Indian Council of Agricultural Research (ICAR), Krishi Vigyan Kendras (KVKs), Agriculture Technology Management Agency (ATMA), Self Help Groups under National Rural Livelihood Mission (NRLM), Farmer Field Schools (FFS), etc. With increase in awareness, well developed drone spraying ecosystem and other sprayers along crop based niche applications and nano fertilisers are expected to grow over time with concomitant reduction and rationalisation in bulk chemical fertilisers consumption.

Widespread adoption of Nano Urea and other nano fertilisers will lead to farmers' profitability and in addressing UN-Sustainable Development Goals (SDGs). It is time that holistic impact of indigenously developed nano fertilisers are understood and encouraged by all stakeholders in unison. The incorporation of nano fertilisers in crop production strategies by the farmers will help in harnessing the benefits of attaining eco-friendly green agriculture as well as food and nutritional security for "Healthy & Fit Bharat".

(This article was published in Rising Kashmir on November 11, 2023)

❑ ❑ ❑

8

Why Rural Bharat Needs Women Drone Pilots

– Dr. Mansukh Mandaviya

Women's empowerment is the key to building a strong and developed nation. This is especially the case when women flourish economically and contribute to rural prosperity. NAMO Drone Didi—an initiative announced by Prime Minister Narendra Modi on November 30 during interaction with beneficiaries in Viksit Bharat Sankalp Yatra—aims to supply drones to 15,000 women Self Help Groups (SHGs) to rent to farmers for agriculture purposes. It is an excellent example of harnessing technological innovation to empower rural women.

This multifaceted scheme effectively addresses the need to modernize our agricultural practices and increase agricultural productivity by placing cutting-edge technology in the hands of rural women. This makes them the epicentre of the rural economy, spearheading the new agricultural revolution. The scheme also opens up new opportunities for the country's young and dynamic start-ups to enter the emerging field of drone aeronautics, which has huge untapped potential.

The emergence of innovative new liquid fertilisers like Nano Urea and Nano DAP with the foliar application has generated the need for the development of an efficient fertigation system. Only the Prime Minister could have envisioned such a brilliant idea—putting drone technology in the hands of women pilots from SHGs.

Though Bharat is the second largest producer of fertilisers, it is also the second largest importer of fertilisers due to inadequate availability of natural resources viz. gas, phosphatic and potassic minerals etc. To overcome this dependence on imports, the government has facilitated revival of several closed down fertiliser units and encouraged setting up of new fertiliser units under "Aatmanirbhar Bharat" scheme. These steps have successfully protected the Bharatiya farmers from the price volatility of fertilisers in the international markets. Through fertiliser subsidies, the Government has been able to ensure that adequate quantities of fertilisers are made available to farmers at the right price at the right time.

Covid-19 pandemic and the recent global geo-political situation posed some tough challenges to the fertiliser sector, and prices soared to unprecedented levels all over the world. We decided to convert these challenges into opportunities. The government encouraged indigenous research for development of alternate fertilisers, the outcome of which is the path breaking liquid nano fertilisers. The next challenge after this new invention was to develop an efficient

Interesting Facts

The need for kisan drones has presented a new range of opportunities for young startups to fulfill their aspirations. Drone manufacturing activity has created immense employment opportunities for the youth of the country. Additionally, this new sector will also create opportunities for rural women pilots, mechanics and spare part dealers.

Do You Know?

This is an excellent example of using technological innovation to empower rural women towards ushering in a new era of rural prosperity. This multidimensional scheme effectively addresses the need to modernize our agricultural practices and increase agricultural productivity by putting cutting-edge technology in the hands of rural women.

fertigation system. That is where the emerging drone technology provided an answer.

Emergence of Kisan-Drones has opened up immense possibilities towards revolutionizing the agricultural practices. The conventional system of spraying pesticides and liquid fertilisers through hand held back strapped pumps is not only time consuming and cumbersome but is also fraught with many dangers from the poisonous reptiles and lurking wild animals in the fields. The automation of the spraying system by means of agri drones provides a time saving and efficient application system. Drone manufacturing activity has created tremendous employment opportunities for the youth. Going further, this new field will also create opportunities for rural women pilots, mechanics and spare part dealers.

I believe that women's empowerment is a prerequisite for achieving "Viksit Bharat". The NAMO Drone Didi Scheme provides new work opportunities to women. It makes them the backbone of the rural economy. This will go a long way in making our agrarian family culture more equitable and strong. A woman drone pilot can play a vital role in increasing the agricultural productivity and efficiency. The application of pesticides and liquid fertilisers through drones will not only save the physical toil and hardship to the farmers but will also be quicker, saving time which can be utilized for more productive work.

The NAMO Drone Didi will definitely be a game changer. It will usher in a new chapter of women empowerment and effectively address multiple issues by providing an opportunity for rural women to be at the centre stage of the economic activity and rural prosperity. There is no doubt that the scheme supported by the newly created network of Pradhan Mantri Krishi Samriddhi Kendras, which provide agricultural services under one roof, will prove to be a harbinger of modern day agricultural revolution and prosperity in a largely agrarian country like ours.

(This article was published in The Indian Express on December 11, 2023)

❑ ❑ ❑

9

The "Drone Moment" of Indian Agriculture

– Dr. Mansukh Mandaviya

Travelling through the verdant serenity of the lush green fields of Punjab, my attention was drawn towards a distant buzzing sound. Curious to know the source, I got down from my vehicle, to be greeted by two rustic farmers operating a drone for spraying liquid Nano Urea. I was pleasantly surprised to see such an enthusiastic acceptance of a new concept by the farmers in a rural corner of the country. This, I felt, was the "drone moment" for the agricultural sector of Bharat, heralding a "drone movement" in our country. The farmers told me that they got information about the agri-drones during their visit to "Viksit Bharat Sankalp Yatra" in their village. The new technology was extremely useful for spraying liquid fertilisers and pesticides most efficiently, they said.

Bharatiya agriculture has already travelled a way by transitioning from being traditionally a bullock cart-based occupation towards tractor-based agriculture. I am now seeing a genesis of third wave of agricultural revolution emerging in the form of usage of drones in agricultural operations. The agri-drone technology is proving to be a real breakthrough in modernizing and transforming our farming practices.

Gone are the days of tedious and time-consuming manual spraying of pesticides and liquid fertilisers through hand pumps; it is gradually getting replaced by a much more efficient and productive technique of spraying by using drones. I always feel that acceptance of any new technology by the rural masses is the key towards spreading the same across India.

Modernization of agricultural practices is essential and indispensable for improving efficiency and productivity of the farming sector so as to ensure food security, which is an important element of Viksit and Aatmanirbhar Bharat.

The green revolution of the 1960s witnessed introduction of new farm tools, HYV seeds, pesticides and chemical fertilisers. Over the years, an urgent need was felt to ensure more balanced use of fertilisers to maintain long term sustainability of our environment, soil health and soil fertility. In response to this need, the Government has embarked upon several initiatives like PM-PRANAM and GOBARdhan to promote alternative fertilisers such as Bio, nano and Organic fertilisers. PMKSK is also set up to enhance awareness about these new initiatives and provide integrated solutions to the farmers under one roof.

Interesting Facts

The development of kisan drones provides an effective and efficient technology for the application of liquid fertilisers. The ability to spray acres of agricultural land in minutes proved to be a boon for farmers who toiled for hours in their fields. With more free time at their disposal, farmers can engage in more productive work to increase their earnings and improve their quality of life.

Do You Know?

Indian agriculture has already moved from being traditionally bullock cart based to being tractor based. I now see the use of drones in agricultural operations as the third wave of agricultural revolution. Agri-drone technology is proving to be a real breakthrough in modernizing and transforming our farming practices.

The indigenously developed nano fertilisers are path-breaking initiative, as they provide an eco-friendly and effective alternative to the conventional chemical fertilisers. A 45 kg bag of conventional urea can be replaced by half-a-litre bottle of Nano Urea which has multifold advantages, like better nutrient use efficiency and easier handling, apart from helping maintain soil quality and improving crop yield. The next challenge in ensuring its wider acceptability was to develop a method of efficient application system, creation of an eco-system for easier adoption and effectively communicating the benefits to the farming community.

The development of Kisan Drones by India's young and dynamic startups offers an effective and efficient technique for application of liquid fertilisers. Having a capacity to cover an acre of farm land within minutes proved to be a blessing to the farmers toiling hard for hours together in their fields. With more free time at their disposal, farmers can engage in more productive work to increase their earnings and improve their quality of life.

Prime Minister Shri Narendra Modi ji, is focussed on empowering the women and increasing the share of women led development within our country. He launched NAMO Drone Didi on November 30, during an interaction with beneficiaries in the Viksit Bharat Sankalp Yatra, which aims to provide drones to 15,000 selected women SHGs for providing rental services to the farmers for agriculture purpose by providing easy financial solutions. This innovative initiative will not only enable easy availability of drones for spraying of liquid fertilisers and pesticides to the farmers, but will also contribute to women empowerment and rural prosperity. It will give a boost to Bharat's drone aeronautics and help drone manufacturing startups to setup manufacturing facilities. It will augment job opportunities for women and youth, creating employment ventures for drone pilots and drone mechanics in the remote corners of the country and will give a fillip to the rural and urban economies at the same time.

Effective communication is the backbone of successful implementation of any novel initiative. Drones have become a star attraction as a part of country-wide Viksit Bharat Sankalp Yatra that started on November 15, 2023. There have been more than 50,000 drone demonstrations across all the States/Union Territories during

> **Interesting Facts**
>
> **Government started various awareness campaign about balanced use of fertilisers like PM-PRANAM, use of nano fertilisers and drone demonstrations as one of the most important component of the Viksit Bharat Sankalp Yatra started from November 15, 2023.**

Do You Know?

Riding on the wings of the drone technology, India's farming sector is marching ahead towards modernization. It takes a strong resolve, clear vision and thorough understanding of the issues, to bring about any new change in the age-old established practices.

the Yatra. These demonstrations have ignited a high level of interest among the farmers at a large scale for adoption of new technology.

The drone movement has given wings to India's farming sector. Yes, I believe we are poised to become the fastest growing economy in the world.

(This article was published in Hindustan Times on December 23, 2023)

❑ ❑ ❑